THE ILLUSTRATED BOOK OF
WILD FLOWERS

ACKNOWLEDGEMENTS

This title first appeared as **The Oxford Book of Wild Flowers** and Peerage Books gratefully acknowledge the co-operation of the Oxford University Press who gave permission for this edition to be published.

The flowers illustrated in this book have been drawn from live specimens, more than half of which were found by the artist herself. The others have been found by many splendid helpers from different parts of the country. We would like to make special acknowledgement to the following:

Mr. E. G. Arthurs, Botany Dept., Oxford University; Miss Janet Burton; Dr. W. A. Clark, F.R.S.E.,F.L.S., Botany Dept., King's College, Durham University; Miss O. Collingwood, All Saints School, Smallridge, Devon; Mrs. D. Fryer; Miss B. Gallois; Miss Dorothy Gray; Mr. J. Hindley, Shaftesbury Modern School; Miss Gwyneth Parry; Dr. Roland Richter, Gordonstoun Science Dept.; Dr. Francis Rose, Bedford College, London University; Miss Mary Wilson; Miss Joanna Wilson; Mr. P. F. Yeo, University Botanic Garden, Cambridge.

We also wish to thank Dr. Lucas, Dept. of Agriculture, Oxford, who has given valuable help in checking the accuracy of the illustrations, and Miss J. Martin, Dept. on Education, Oxford, who has helped to prepare the pages on Flower Families.

The majority of the text drawings have been prepared by Eileen Green, and the remainder by Joan Sampson. They had valuable help in finding material from the Herbarium, Dept. of Botany, University of Oxford.

In naming the flowers in this book, we have followed Clapham, Tutin and Warburg's *Flora of the British Isles* (Cambridge 1952). Where more than one English name is in common use, we have given both.

THE ILLUSTRATED BOOK
OF WILD FLOWERS

Illustrations by

B. E. NICHOLSON

Text by

S. ARY & M. GREGORY

PEERAGE
BOOKS

First published in Great Britain by Oxford University Press as
The Oxford Book of Wild Flowers

This edition published by Peerage Books
59 Grosvenor Street
London W1

© 1960 Oxford University Press
1980 Revised
Reprinted 1984, 1985

ISBN 0 907408 49 4

Printed in Czechoslovakia
50535/2

Contents

INTRODUCTION

This book is designed to help those with little or no botanical training who wish to identify the wild flowers they see in this country. Since the most conspicuous feature of a flower is usually its colour, the plants are arranged here by colour rather than by family. This is an arrangement for convenient identification and cuts across botanical classification; but it has often been possible to group plants of the same family on one page, and in any case the family to which a plant belongs is always mentioned.

It is probable that, as more and more plants are identified, and their families noted, the characteristics of at least some of the families will become well known. For those who wish to learn more about the classification of flowering plants, there is a section on this subject, beginning on page 209, which gives brief descriptions of the more important characteristics of the main flower families, and lists of those species of each family illustrated in this book. Familiarity with common plants and the easy introduction to classification provided here will soon enable the user of this book to be able to make use of a more technical book to identify those rarer plants which lack of space prevented our being able to include.

550 species are illustrated in full colour and about 30 more in black and white. Other species, usually less common ones, are described in the text together with those related species which are illustrated and which they most closely resemble.

Space has made it impossible to treat grasses and sedges, though both are flowering plants. They are large groups and would have needed much space to have been presented adequately.

This is how to identify a plant. If the flower is yellow, for example, first look through the yellow plates to find the picture most like your specimen. Then read the description to see if it confirms your guess. Points to note from the picture are leaf shape and colour, flower shape and colour, and the number and arrangement of the sepals, petals, stamens, and carpels. Many smaller flowers may often be more easily examined by using a magnifying glass. In the description notice especially the size of the plant: most flowers, but not all, are portrayed life-size in the illustrations. Note also when the plant usually flowers, in what part of the country it grows, on what kind of soil it flourishes most, and whether it is usually found in woods, marshes, grasslands, etc. If you have correctly identified your flower, all these details should tally with your specimen. The flower colour often varies from one specimen to another, and so does size, so the flower illustrated may not exactly match your specimen. Details of leaves, sepals, petals, etc. should not vary unless the text tells you they do; and the text will also mention any outstanding variations in colour and size.

Several thousand species of flowering plants have been recorded in the British Isles, many of which are extremely rare. You may find one of these rarer plants which is not covered in this book. This book should, however, give you an idea what kind of flower it is likely to be, and then you will have to resort to a more detailed book to identify it. In this book the scientific Latin name is given immediately following the common name and both are included in the index. In a more detailed 'flora' the scientific name only may be used. More details about the importance and meaning of scientific names will be found on page 206.

Some plants, especially the common weeds, may be found almost anywhere; some, however, are found only in specialized habitats. Some plants, for example, will grow only on limestone or only in dry, sandy soils; others will not grow at all on limestone or need heavy, wet soils. These facts are a help in identification, and more is said about this in a section on page 220.

Many people, when they find wild flowers, like to pick them. This may be simply for the pleasure of picking or it may be to add to a collection of pressed flowers or to be able to paint the flower at home. There is an important thing to be remembered about picking flowers. Flowers produce fruit containing seeds. When the fruit is fully ripe, the seeds are dispersed, and from these seeds come future plants. In order to keep the supply of plants going, a great many seeds must be produced, since there is an inevitable wastage. If all the flowers are picked, there can be no seeds at all. In the case of annuals and biennials, that will mean no flowers at all next year. In the case of perennials, the old plant will flower again, but if the picking is repeated, in time the old plant dies and there are no new ones to replace it.

Picking wild flowers must be done with discretion and always with an eye to next year's crop. These are good rules. Never pick except for a good purpose. If there are only one or two flowers, do not pick them on any account at all. If there are three, pick one if you really need it. Always leave behind two or three for every one you pick. If you find rare wild flowers, do not tell other people where they are unless they are people to be trusted. Note down in a nature diary where you have found the flowers, and look again next year to see if they are still there. A nature diary showing what flowers you have found on what dates and where is an interesting thing to keep. As you grow more interested in wild flowers from using this book, you will become more interested in preserving our wonderful heritage of wild flowers for the future.

GLOSSARY

These are words used in the text and which the reader is expected to understand.

Achene. A dry, single-seeded fruit formed from one carpel and indehiscent — that is, it does not split to release the seed (Fig. 3).

Annual. A plant completing its life cycle within one year and flowering once only.

Anther. The part of a stamen in which the pollen grains are produced (Fig. 1).

Biennial. A plant completing its life cycle in two years, flowering only in the second year.

Bract. A small modified leaf at the base of a flower stalk or immediately below a flower-head (p. 32). It is usually green and leaf-like, but sometimes scaly.

Calyx. The outermost part of the flower, consisting of the sepals, which serves to protect the rest of the flower bud. It is usually green, and the sepals are sometimes separate and sometimes joined to form a tube (Fig. 1).

Capsule. A dry fruit made of more than one carpel and dehiscent — that is, it splits to release the seeds (Fig. 3).

Carpel. One or more units, each consisting of stigma, style, and ovary, of which the female part of the flower, or pistil, is built up.

Corolla. The petals all together form the corolla, a word most often used when the petals are coloured and joined at the base (Fig. 1).

Filament. The stalk of the stamen (Fig. 1).

Follicle. A dry, dehiscent (splitting) fruit formed from one carpel and containing several seeds. It differs from a pod or legume because, when ripe, it splits only along the inner seam of the fruit (Fig. 3).

Involucre. Row or rows of bracts below a flower-head (p. 36).

Node. The point on the stem, often thickened, from which a leaf springs (Fig. 2).

Ovary. The basal part of one or more carpels containing the ovules, in which the female eggs are produced, which eventually become the seeds (Fig. 1).

Pappus. The downy hairs to be seen on some fruits, such as Thistles and Dandelions (pp. 40, 150, etc.), which form the calyx.

Perennial. A plant which lives for years, usually flowering each summer after reaching maturity.

Perianth. The sepals and petals together — a word most often used when the sepals and petals are similar.

Petals. These are usually brightly coloured and may serve to attract insects (Fig. 1). In some flowers they are absent or not developed.

Pistil. The whole of the female part of the flower consisting of one or more carpels.

Pod or legume. A dry dehiscent (splitting) fruit formed from one carpel and differing from a follicle in that, when ripe, it splits along both inner and outer seams (Fig. 3).

Pollen. Small grains produced in the anthers, each grain containing a male cell necessary for fertilizing the egg.

Receptacle. The swollen head of the flower-stalk which carries the different parts of the flower (Fig. 1). In some of the Rose family the receptacle forms a cup enclosing the female parts of the flower (pp. 79, 117).

Rhizome. An underground rooting stem growing horizontally just beneath the soil surface and persisting for more than one growing season.

Sepals. The outer parts of the flower, usually green, which form the calyx (Fig. 1).

Stamen. The male part of the flower consisting of a filament bearing a pollen-producing anther (Fig. 1).

Stigma. The top of the carpel where the pollen grains are received. It is often sticky and knobbly so that the pollen sticks to it (Fig. 1).

Stipules. Two small, often leaf-like structures at the base of the leaf stalk (Fig. 2).

Style. The part of the carpel between the stigma and the ovary (Fig. 1).

Tuber. A swollen underground rooting stem, such as the potato, lasting only one year. It is also sometimes a swollen root, as in the Lesser Celandine (p. 5).

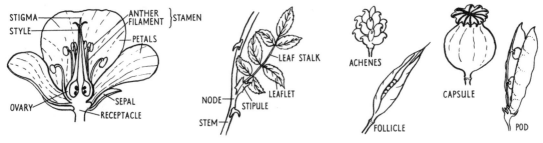

Fig. 1   Fig. 2   Fig. 3

1 **Bulbous Buttercup** (*Ranunculus bulbosus*). The distinguishing characteristics of this species are the swollen, bulb-like base of the stem (hence its Latin and English names), the hairy, furrowed flower stalks, and the way in which the sepals turn back against the flower stalk when the flower opens. As with some other yellow flowers, the flowers appear a more orange-yellow than they actually are because the orange colour of the pollen is reflected in the shiny upper surface of the petals. The fruits are achenes, as are those of all Buttercups, but in this case they are smooth. The Bulbous Buttercup is common in dry meadows throughout the British Isles and flowers earlier than the other common Buttercups. (*April—June*).

Hairy Buttercup (*R. sardous*). An annual species which looks rather like a small Bulbous Buttercup but has paler flowers, paler and more hairy leaves, and is without the bulbous stem base. It is much less common, and is most often found near southern coasts; less frequently in the north. (*May—July*).

Creeping Buttercup (*R. repens*). This common species is rather like a late-flowering Bulbous Buttercup, but has larger leaves, erect sepals, and no bulbous stem base. It can be a tiresome weed, spreading rapidly by leafy runners (Fig. 1). These put out roots from the point where the leaves join the stem, thus starting new plants. It is especially common on damp, heavy soils. (*May—September*).

Fig. 1

2 **Small-flowered Buttercup** (*R. parviflorus*). A rather uncommon annual, found locally on bare, dry ground, especially on chalk and limestone. It is a pale green, hairy, sprawling plant, with pale flowers much smaller than those of any other Buttercup. It is difficult to recognise as a Buttercup, but in fact it has the typical characteristics of numerous stamens and carpels. The flower stalk is furrowed, and the fruits are very rough. (*May—July*).

3 **Meadow Buttercup** (*R. acris*). This is the tallest of the common Buttercups, growing up to 3 feet in height. It has the typical Buttercup leaves divided into three lobes, but in this species the middle lobe is not stalked. The flower stalks are smooth and carry more flowers on a stem than do most Buttercups. The Latin name means that the plant has a bitter taste, as have most Buttercups, and cattle will not usually eat it. It is common throughout the British Isles in damp meadows, pastures, and waysides. (*May—July*).

Corn Buttercup or Corn Crowfoot (*R. arvensis*). A small, hairy, annual plant, differing from other Buttercups in two ways: the lobes of its leaves are narrower, rather like the upper leaves of the Meadow Buttercup; and its fruits are covered with long spines (Fig. 2) These spiny seeds made it a tiresome weed in cornfields, and have earned it some of its local names, for instance, Devil's Claws, Devil-on-all-Sides, and Scratch Bur. It used to be common in cornfields and waste places, but now that more weed-killers are used and seed corn is cleaned more thoroughly, it is quite rare. (*June—July*).

Fig. 2

4 **Goldilocks** (*R. auricomus*). This is the only British woodland Buttercup and the only Buttercup without a bitter taste. The flowers often have a ragged, bird-pecked appearance because some of the petals may be tiny or missing altogether. The flowers may have any number up to 7 petals. It is found in suitable places throughout the British Isles. (*April—May*).

5 **Lesser Spearwort** (*R. flammula*). This species differs from other Buttercups in having long, rather narrow leaves, the lower leaves being broader and less spear-like than the upper. It is a common plant of wet places such as ditches and marshes, especially in the north and west, and may grow up to 2 feet tall. Some Spearwort plants have slender, creeping stems, rooting at irregular intervals, and these grow less tall and usually have narrower leaves and smaller flowers. (*June—August*).

Greater Spearwort (*R. lingua*). This is in general like a very large Lesser Spearwort, with the same long spear-like leaves. It is most easily distinguished from the Lesser Spearwort by the size of its flowers, which are usually more than twice as large — perhaps up to 2 inches across. The fruit has a curved beak. It is an uncommon plant of marshes and fens. (*June—August*).

LIFE SIZE

1 BULBOUS BUTTERCUP   2 SMALL-FLOWERED BUTTERCUP   3 MEADOW BUTTERCUP
4 GOLDILOCKS   5 LESSER SPEARWORT

RANUNCULACEAE: BUTTERCUP FAMILY (See p. 209)

1 Celery-leaved Crowfoot (*Ranunculus sceleratus*). This smallest-flowered of the Buttercups grows in wet places, as do the Spearworts (p. 2), but can be distinguished from them not only by its small flowers but also by the difference in shape of its lower and upper leaves. The plant grows from 6 to 18 inches tall, with hollow stems which contain a very bitter juice; this may cause blisters and ulcers if it is rubbed on the skin. Beggars once used it to bring up blisters which they hoped would excite sympathy and bring them money. This may be the reason for the name *sceleratus*, which means wicked, or vicious. The plant is common throughout the British Isles, especially in rich mud. (*May—September*).

There is another species in the Scilly Isles, *R. muricatus*, which is shorter and bushier and has larger flowers. It is a tiresome weed in bulb fields.

2 Mouse-tail (*Myosurus minimus*). Both the English and Latin names of this plant refer to the supposed resemblance of its fruiting receptacle to a mouse's tail. It is a plantain-like stalk covered with tiny brown achenes. The plant grows from 2 to 4 inches tall and has rather grass-like leaves and minute greenish flowers. It grows in low-lying sandy places in England and Wales and is rather rare. (*April—July*).

3 Globe Flower (*Trollius europaeus*). The flowers, larger than those of any other Buttercup, are 1 inch across and have no green sepals. Inside the flower are 5 to 15 yellow tubular nectaries. The fruit consists of a group of follicles, each containing several small black seeds (Fig. 1). Globe Flowers are found locally but not very commonly in damp pastures and woods in the hilly parts from Wales northwards and in north-west Ireland. (*May—August*).

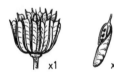

x1   x1

Fig. 1

4 Common Meadow Rue (*Thalictrum flavum*). The flowers are borne in compact clusters, each flower having 4 whitish petals which fall very soon after the flowers open. The yellow colour is given by the numerous feathery stamens, characteristic of the family, which are held erect. The plant grows from 2 to 4 feet tall, and is fairly common in wet meadows and by streams throughout Great Britain, except in the extreme north. (*June—August*).

Alpine Meadow Rue (*T. alpinum*). This rather insignificant little mountain plant, rarely exceeding 6 inches in height, grows principally on moist rocky ledges and turf in North Wales and northwards from Yorkshire. It has dark green, fern-like leaves, and tiny flowers borne on a single 3-inch spike. The yellow and purplish stamens which give it its colour hang down instead of being held erect. (*May—July*).

Lesser Meadow Rue (*T. minus*). This plant, though it may grow as tall as Common Meadow Rue, is much more slender and wiry, with smaller, more delicate leaves. The flowers are not borne in compact groups, and the more greenish-yellow stamens hang down instead of standing erect. Lesser Meadow Rue is locally common in a variety of places, chiefly in the north. (*June—August*).

5 Marsh Marigold (*Caltha palustris*). The name Marigold is misleading, and its origin unknown. The plant has many other names such as Kingcup and May Blobs. The flowers have no green sepals. The lower leaves are stalked. The stout hollow stems grow up to 18 inches tall. They are held erect when the plant is growing in lowland marshes, but in the north and on mountains smaller plants with trailing slender stems are more usual. (*March—June*).

6 Lesser Celandine (*Ranunculus ficaria*). This plant and the unrelated Greater Celandine (p. 6) may both have been named from the Greek word for swallow, *chelidon*, because they flower when the swallows return. In fact the Lesser Celandine flowers some 2 or 3 months ahead of the swallows. Some Lesser Celandine plants have little swollen buds, called bulbils, by which the plants spread; those plants producing bulbils rarely produce seed. It is common, especially in damp, shady places, and can be a tiresome weed. (*March—May*).

LIFE SIZE

1 CELERY-LEAVED CROWFOOT 2 MOUSE-TAIL 3 GLOBE FLOWER

4 COMMON MEADOW RUE 5 MARSH MARIGOLD 6 LESSER CELANDINE

1 Yellow Horned Poppy (*Glaucium flavum:* family *Papaveraceae*). This seaside plant usually grows on shingle and may be found all around the coast of the British Isles except in the north of Scotland. It is rarely found inland in Britain but more frequently so on the continent. It grows from 1 to 2 feet high, and the seed pods may reach 6 to 12 inches long when fully ripe. A sticky yellow juice oozes out when the bluish-green stems or leaves are broken. (*June—September*).

2 Welsh Poppy (*Meconopsis cambrica*). A delicate little Poppy, growing from 1 to 2 feet tall. It is most often found in damp, rocky places in Wales and north-west England. This is the only species of *Meconopsis* found wild in Western Europe; all the others, grown frequently in gardens, and sometimes with large blue flowers, are native to Asia, especially to the Himalayan region. Both the yellow and an orange form of the Welsh Poppy are also grown in gardens and are sometimes found as garden escapes. It has a yellow juice and long, ribbed capsules (Fig. 1). (*June—August*).

x1

Fig. 1

3 Greater Celandine (*Chelidonium majus*). This plant has no connection with the Lesser Celandine (p. 4) except in name. The bright orange juice in its stems, which is poisonous, used to be used for treating warts and eye complaints. In fact, it may have been cultivated for this purpose, and it is still most frequently found on banks and in hedges near houses. The seeds carry whitish fleshy appendages which ants use for food and, in the course of collecting them, the ants distribute the seeds (Fig. 2). The plant grows from 12 to 18 inches tall. (*May—September*).

4 Wild Pansy or Heartsease (*Viola tricolor:* family *Violaceae*). The yellow-flowered Heartsease is not un-

common, though the purple and yellow variety is more frequently found. This plant is described in detail on page 164, in the purple section.

5 Yellow Waterlily (*Nuphar lutea:* family *Nymphaeaceae*). This well-known Waterlily usually grows in still or very slow-moving water, with its wide, leathery leaves floating on the surface. It may sometimes grow in faster-moving water, in which case the leaves become submerged and are thin and cabbage-like. The flowers are held a few inches above the surface of the water and smell like stale wine. This and the shape of the fruits may explain its other name — Brandy-bottle. (*June—September*).

Least Yellow Waterlily (*N. pumila*). This very rare Waterlily is smaller in every way than the common Yellow Waterlily, and is found only in lakes in Scotland, Shropshire, and Merioneth. Besides being smaller, it has wider gaps between the petals. (*July—August*).

Fringed Waterlily (*Nymphoides peltatum:* family *Menyanthaceae*). A plant of quiet waters especially disused canals, mainly in the south of England. It is becoming more frequent than it used to be. Its leaves, smaller than those of the Yellow Waterlily, float on the water and are purple underneath and sometimes purple spotted on top. The flowers have 5 yellow, petal-like corolla lobes with distinctly fringed edges. Several flowers grow from the main stem. (*July—August*.)

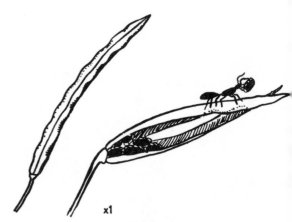

x1

Fig. 2

LIFE SIZE

1 Yellow Horned Poppy   2 Welsh Poppy   3 Greater Celandine
4 Wild Pansy   5 Yellow Waterlily

CRUCIFERAE: CABBAGE FAMILY (See p. 209)

1 Black Mustard (*Brassica nigra*). This plant has been cultivated for centuries because both mustard and an oil used in medicine and soap-making are obtained from its seeds. It also grows wild on sea cliffs, stream banks, and waste places, and is fairly common except in north Scotland and Ireland. It grows from 2 to 3 feet tall and has stalked leaves, the lower ones being bristly, while the upper are narrow and smooth. The ripe seed-pods are held close to the stem. (*May—August*).

Hoary Mustard (*Hirschfeldia incana*). An annual plant, like Black Mustard but smaller and paler and densely covered with coarse whitish hairs, especially on the lower leaves and stem. The pods have a one-seeded beak (Fig. 1a). The leaves are lobed and toothed. It is a native of Mediterranean countries but is becoming established in a few waste places in southern England. (*May—September*).

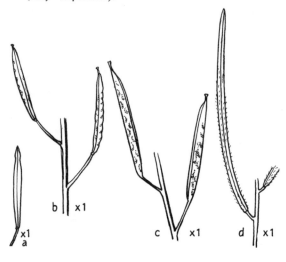

Fig. 1

2 White Mustard (*Sinapis alba*). This is the mustard of 'mustard and cress'. It is cultivated both as a fodder crop and for the mustard obtained from its seeds. It can be distinguished from Black Mustard by the whitish hairs on its pods, which are held away from the stalk, and from Charlock by its more divided leaves. It is a common weed throughout Great Britain, especially on chalky soils. (*May—August*).

3 Wall Rocket (*Diplotaxis tenuifolia*). This and Stinkweed are the only yellow-flowered Cruciferae to have an unpleasant smell when crushed. Wall Rocket is usually larger (1 to 2 feet) and bushier than Stinkweed. Its hairless leaves grow near the base of the stem which is smooth, leafless, and rather woody. The pods, which are about an inch long, are separated from the flower stalk by a distinct constriction (Fig. 1b). It is fairly common on walls, railway banks, and waste places, especially on chalk in southern England, but is rare further north. (*May—September*).

Stinkweed (*D. muralis*). A close relative of Wall Rocket described above. It has fewer, smaller, and paler flowers, and its leaves grow mostly in a rosette. The pods are without the distinct constriction (Fig. 1c). It is rather more widespread, especially on southern sandy soils. (*May—September*).

4 Hedge Mustard (*Sisymbrium officinale*). This annual, 1 to 2 feet high, can be recognised by the characteristic growth of the branches, almost at right angles to the main stem. The small pods are pressed tightly to the stem. It is a very common weed and is found throughout Great Britain. (*May—July*).

Eastern Rocket (*S. orientale*). A native of southern Europe and the Near East which has established itself on waste ground in southern England. It appeared frequently on bombed areas in London. Its growth is loose and variable. The lower leaves form a rosette and have rather narrow lobes, while the upper are spear-like, almost without lobes. The seed-pods, which are hairy when young, grow from 2 to 3 inches long (Fig. 1d). (*May—August*).

Another closely related species, **Tumbling Mustard** (*S. altissimum*), is a spreading, slender annual most easily distinguished from Eastern Rocket by its leaves, which are all pinnate, with very narrow lobes. It is usually taller and has no hairs on its very long pods. It is locally frequent in waste places. (*June—August*).

5 Treacle Mustard (*Erysimum cheiranthoides*). This Mustard is distinguished by its branched and rather star-shaped hairs, its long, four-angled, slightly curved seed-pods, and its spear-shaped, barely toothed leaves. It is a slender annual, growing 6 inches to 2 feet high, and fairly common on low-lying waste ground in southern England. It is rare in the north. (*June—August*).

6 Charlock (*Sinapis arvensis*). One of the most troublesome weeds of cornfields and other cultivated ground. Its seeds can survive for long periods buried in the soil and can germinate when brought to the surface by ploughing after 12 years or possibly much longer. Selective weed-killers are beginning to get it under control, although it is still abundant in waste places. It has a coarse growth and is usually bristly, at least on the lower parts. The fruits are held away from the stem and have long beaks. (*May—July*).

LIFE SIZE

1 BLACK MUSTARD   2 WHITE MUSTARD   3 WALL ROCKET

4 HEDGE MUSTARD   5 TREACLE MUSTARD   6 CHARLOCK

9

CRUCIFERAE: CABBAGE FAMILY (See p.209)

1 **Woad** (*Isatis tinctoria*). The hanging purple seed-pods of Woad distinguish it from all other British Cruciferae. It used to be used as a dye and was cultivated in Britain for more than 1000 years. To obtain the dye, the leaves were dried, powdered, and fermented — a tedious and very smelly process. The last woad mills in the world, in Lincolnshire, were closed during the 1930's. Woad can be easily grown in gardens, but it is now rarely found wild except on cliffs near Tewkesbury and Guildford. It is a biennial, growing from 18 inches to 3 feet high. (*June—August*).

2 **Common Winter Cress** or Yellow Rocket (*Barbarea vulgaris*). This stocky, dark green, hairless plant has a basal rosette of stalked leaves which remain green all winter. Its bitter taste, however, makes it useless for salads. The flowering stem grows a foot or more high and carries unstalked leaves. It is common in damp places in England, but less common in Scotland. (*May—June*).

An early flowering and much less common relative is **Early Winter Cress** (*B. intermedia*), which has smaller and more deeply divided stem leaves (Fig. 1*a*), smaller flowers, and rather longer seed-pods. It grows in waste places and fields. (*April—May*).

Fig. 1

Land-cress (*B. verna*). The leaves of this plant do not taste bitter, and it used to be grown as a salad plant, especially in south-west England. It has larger flowers than Common Winter Cress, with petals about three times as long as the sepals. Its stem leaves are much narrower and rather paler (Fig. 1*b*), and its pods longer and more spreading. In spite of its name it is probably a native of western Mediterranean countries. (*April—June*). There is also a Small-flowered Land-cress (*B. stricta*), a slender plant with less toothed upper leaves. It is very rare.

3 **Marsh Yellow-cress** (*Rorippa islandica*). An erect annual plant of pond sides and river banks, especially where water stands during the winter; it does also grow in dry places such as railway banks. The stem is hollow and grows up to 2 feet tall. The petals are the same length as the sepals, which spread outwards when the flower opens, and the seed-pods are rather 'bottle-shaped' (Fig. 2*a*). (*June—September*).

Creeping Yellow-cress (*R. sylvestris*). This plant can be distinguished from Marsh Yellow-cress by its more slender, upward-pointing pods (Fig. 2*b*), its petals which are longer than the sepals, its tattered-looking, paler green leaves, and its more sprawling growth. It has creeping, rooting stems by which it spreads. Although a little less common than Marsh Yellow-cress, it is still quite frequent in damp places, and is sometimes a troublesome weed in gardens. (*June—August*).

Greater Yellow-cress (*R. amphibia*). This Yellow-cress is much larger than the other two species, reaching 2 to 4 feet in height. Its leaves are toothed but not deeply lobed, and its petals are twice as long as the sepals. The small egg-shaped fruits are carried on long stalks (Fig. 2*c*). It grows by water, frequently in the south but rarely in the north. (*June—September*).

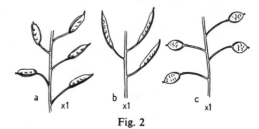

Fig. 2

4 **Flixweed** (*Descurainia sophia*). This plant is distinguished from all others in the family by its very finely divided leaves and bushy growth. It may reach 3 feet in height. The flowers are tiny, with petals and sepals of the same length, but the narrow fruits are surprisingly long, up to one inch. It is widespread in bare, waste places, but nowhere very common. (*June—August*).

5 **Field Cabbage** (*Brassica campestris*). An annual wild form of the biennial Turnip, which has a less swollen tap root. It grows from 1 to 3 feet high, and differs from many other members of the family in that its flowers are bunched together because the main stalk does not elongate as the flowers open. It is frequent on bare ground, especially as a relic of cultivation. Rape and Swede are other forms which have been developed as field crops. (*April—August*).

LIFE SIZE

1 WOAD   2 COMMON WINTER CRESS   3 MARSH YELLOW-CRESS

4 FLIXWEED   5 FIELD CABBAGE

11

1 Common St. John's Wort (*Hypericum perforatum:* family *Hypericaceae*). The Latin name, *perforatum*, refers to a number of small transparent dots in the leaves of this plant which are groups of oil-producing cells. The English name is because in many countries the St. John's Worts were associated with St. John the Baptist. The word 'wort' means plant or herb. This species was used for treating wounds and was hung up in doors and windows to prevent the entry of evil spirits. The distinguishing features of Common St. John's Wort are the two raised lines on the stem and the black dots on the petals and sepals. The many stamens are grouped into three bundles. The plant is common in grassland and open scrub, except in the north, and grows 1 to 2 feet high. (*July—September*).

Square St. John's Wort (*H. tetrapterum*). This also common species is distinguished by its square stems with winged angles (Fig. 1). Its leaves are broader than Common St. John's Wort, and its flowers are paler and only about half the size. It is found in damp places throughout the British Isles. (*July—September*). There is another less common species, *H. dubium*, called Imperforate St. John's Wort because the leaves are usually without the transparent dots. Its flowers are larger and a more golden yellow. It prefers damp, shady places and rather heavy soils. (*June—August*).

Fig. 1

Hairy St. John's Wort (*H. hirsutum*). The hairy stems and leaves of this plant distinguish it from the other species. Its flowers are paler yellow than Common St. John's Wort, and it has black dots on the sepals only. It is common in dry, shady places, especially on chalk, limestone, and clay. (*July—September*).

2 Slender St. John's Wort (*H. pulchrum*). This slender plant, growing 9 to 18 inches tall, has flowers tinged with red underneath, smooth stems, and almost heart-shaped leaves growing in pairs. It is widespread in heathy places, but not on chalk or limestone. (*July—August*).

Mountain St. John's Wort (*H. montanum*). An upright plant with few or no branches and only slightly hairy.

The leaves have no transparent dots, but a row of black dots on the back. There are small clusters of pale flowers. The species is found in some shady places, usually on chalk or limestone. (*July—September*).

3 Trailing St. John's Wort (*H. humifusum*). Its name indicates its trailing habit. It has slender almost wiry stems, small, pale leaves, often with black dots, and small flowers along the stem. The upper leaves usually have a few transparent dots. It is fairly common in heathy places and open woodland. (*June — September*). A larger, stouter, and more erect species, Flax-leaved St. John's Wort (*H. linarifolium*), is occasionally found on dry, rocky places in the south-west.

4 Wild Parsnip (*Pastinaca sativa:* family *Umbelliferae*). The Wild Parsnip has a strong, very unpleasant smell, which attracts flies and beetles to the flower. It is a stout, hairy plant, growing from 2 to 4 feet tall. It is fairly common throughout the British Isles, especially on chalk and limestone in the south and east. (*June—September*).

5 Pepper Saxifrage (*Silaum silaus:* family *Umbelliferae*). Since this plant is neither a Saxifrage nor is it peppery, its name is rather surprising and the reason for it is not known. It is a stiff plant, growing 1 to 2 feet high. Its finely divided lower leaves have rough edges, and the fruits are egg-shaped. It grows best on damp, heavy soils, and is fairly common in meadows and on roadsides in the south. It is rarer in the north and not found at all in Ireland. (*June—September*).

Parsley (*Petroselinum crispum*). This well-known garden herb also grows wild in some places, especially near the sea. Its groups of greenish-yellow flowers grow on strong stems 9 to 18 inches tall. (*June—August*).

6 Orange Balsam or Jewel Weed (*Impatiens capensis:* family *Balsaminaceae*). All the Balsams are easy to recognise by the flower's broad lower lip, small upper hood, and spur behind. The name *impatiens* refers to the fact that the ripe seeds explode when touched, shooting out their seeds. The rather rare Yellow Balsam of the north-west (*I. noli-tangere*) carries the English name Touch-me-not for this reason. Jewel Weed came originally from eastern North America, and is now increasing on river banks in south-eastern counties. It is a bushy plant growing 3 to 4 feet high. Small Balsam (*I. parviflora*) grows only ½ to 2 feet high, and has small, pale, un-spotted flowers. It is found locally in dry shady places. (*July—September*).

THREE-QUARTERS LIFE SIZE

1 COMMON ST. JOHN'S WORT   2 SLENDER ST. JOHN'S WORT   3 TRAILING ST. JOHN'S WORT

4 WILD PARSNIP   5 PEPPER SAXIFRAGE   6 ORANGE BALSAM

1 **Golden Saxifrage** (*Chrysosplenium oppositifolium:* family *Saxifragaceae*). A little plant of wet, shady places which often forms dense mats of creeping, rooting stems bearing leaves in opposite pairs. The upper stem-leaves form a rosette beneath the groups of tiny 4-sepalled, petal-less flowers. It is fairly common in suitable places, especially in the north and west of the British Isles. (*March—May*).

Alternate Golden Saxifrage (*C. alternifolium*). This Golden Saxifrage can be distinguished from the last species by its leaves, which are longer stalked and alternate, not opposite. It is a bigger plant, does not form mats, and has longer flower-heads. It often grows with Golden Saxifrage but prefers an even shadier place, especially under overhanging rocks. It is less common. (*March—May*).

2 **Bog Asphodel** (*Narthecium ossifragum:* family *Liliaceae*). The Latin name of this plant, *ossifragum*, means bone-breaking, and it was once believed that the bones of cattle feeding on it became brittle. The yellow, starry flowers turn to a bright, deep orange as the fruit ripens, giving the impression that the plant is still flowering late into the autumn. It grows about 6 inches high and is widespread in bogs and wet heaths on acid soils, being more common in the north and west than the south. (*July—August*).

3 **Roseroot** (*Sedum rosea:* family *Crassulaceae*). The thick, fleshy roots of Roseroot, when cut, give off a

Fig. 1

scent like a rose scent — hence its name. The male and female flowers grow on different plants, the male flowers having longer petals than sepals, and the female having both the same length (Fig. 1). The fruits when ripe are orange and may easily be mistaken for flowers. Roseroot grows 6 to 12 inches tall and is found on mountain ledges in Wales and north England and sea cliffs in west Scotland and Ireland. (*May—June*).

4 **Honeysuckle** (*Lonicera periclymenum:* family *Caprifoliaceae*). Its other name, Woodbine, refers to its habit of twining tightly round the trees and shrubs on which it climbs. Honeysuckle flowers first open at twilight and are then white or pale pink. Their light colour and strong scent attract moths whose long tongues can reach the nectar at the base of the tube and who usually pollinate the flowers as they hover in front of them while feeding. Bees sometimes steal nectar without pollinating the flower by biting through the base of the tube. The flowers turn orange-brown after pollination. (*June—September*).

Fly Honeysuckle (*L. xylosteum*). This is a much rarer species, though often grown in shrubberies. It is an upright shrub and has unscented flowers growing in pairs at the base of opposite leaves. It is best known in Sussex. (*May—June*).

5 **Yellow Stonecrop** (*Sedum acre:* family *Crassulaceae*). The Latin name, *acre*, meaning bitter, and its other English names, Wall Pepper and Biting Stonecrop, refer to the bitter, peppery taste of its tiny swollen leaves. It is the commonest and smallest of the Yellow Stonecrops. It forms mats of short stems, 2 to 4 inches tall, in dry places such as walls, roofs, sand dunes, and dry grassland throughout the British Isles. (*June—July*). For other *Sedums* see pp. 83, 113.

Rock Stonecrop (*S. forsterianum*). This *Sedum* looks like an elongated Yellow Stonecrop but with longer leaves, often pink tinged, which remain on the stem after they die. The flowering stems grow erect and reach 6 to 12 inches in height. It is a rare plant, growing in rocky places in the west of England and Wales. (*June—July*).

Large Yellow Stonecrop (*S. reflexum*). This *Sedum* is commonly cultivated in gardens, in old days as a salad plant, and sometimes escapes from gardens on to walls and rocks. It is larger and stouter than Rock Stonecrop, and its greyish-green leaves, sometimes tinged with pink, drop off when they die. The flowers are clustered in heads which hang down until the flowers open. (*July—August*).

6 **Rock Rose** (*Helianthemum chamaecistus:* family *Cistaceae*). A common wild flower throughout the British Isles in grassy places on chalk and limestone, and also in gardens with many different coloured varieties. It is not a Rose at all, and it was given its Latin name, meaning sun flower, probably because the flowers open fully only in bright sunshine. The leaves are thickly covered underneath with white hairs. (*May—September*).

The rare White Rock Rose (*H. apenninum*), found only on limestone near Weston-super-Mare and Torquay, is like a large common Rock Rose except that its flowers are white.

Hairy Rock Rose (*H. canum*). This rare plant of limestone rocks in Wales and north England (although common round Galway Bay in Ireland) is a smaller and less straggly Rock Rose. The silvery leaves are usually hairy all over, and the flowers are only about half the size of those in the common Rock Rose. (*May—June*).

1 GOLDEN SAXIFRAGE   2 BOG ASPHODEL   3 ROSEROOT

4 HONEYSUCKLE   5 YELLOW STONECROP   6 ROCK ROSE

LIFE SIZE

ROSACEAE: ROSE FAMILY (see p. 212)

1 Herb Bennet or Wood Avens (*Geum urbanum*). This plant is not, in fact, confined to woods but is also found in shady, rather damp places throughout the British Isles. The green lobes alternating with the sepals, called the epicalyx, are characteristic of the Rose family. The little hooks on each fruit help to distribute the seeds by catching in the fur of passing animals or in birds' feathers. The roots, which have a sweet, clove-like smell, were once thought to have the power of protecting a house against the devil. This plant sometimes hybridizes with Water Avens (p. 117) when the two are growing together. (*June—August*).

2 Silverweed (*Potentilla anserina*). The plant may have got its Latin name because it grows freely on closely grazed grass such as goose greens (*anser* means a goose). It spreads by creeping rooted stems. It once had many uses: it was used to treat ulcers and sores, and was often placed in shoes to keep the feet comfortable when walking long distances. When food was short, the roots were often eaten, raw, boiled, roasted, or ground into a mealy flour. Silverweed is widespread in damp grassy and waste places. (*May—August*).

3 Shrubby Cinquefoil (*P. fruticosa*). Its greyish-green, slender, divided leaves and shrubby growth distinguish this plant from all other Cinquefoils. The bush grows 1½ to 3 feet tall. It is found wild only on limestone rock in the Lake District, Upper Teesdale, and west Ireland; but it and similar species are frequently grown in gardens. (*May—July*).

4 Tormentil (*P. erecta*). The roots of Tormentil used to be boiled in milk to treat diarrhoea in children and calves. They were also used in the north as a substitute for oak bark in tanning, and a red dye was obtained from them. Its leaves are divided into 3 toothed leaflets with 2 basal leaf-like stipules, which give a five-fingered effect. The flowers almost always have 4 sepals and petals, instead of the more usual 5. Tormentil has a prostrate growth, but its slender stems never form roots. It is abundant on acid soils on moors, heaths, and grassy places. (*May—September*).

A similar but much rarer Potentilla is Hoary Cinquefoil (*P. argentea*), which grows in dry sandy and gravelly soils. It has smaller flowers with 5 petals and sepals, and the stem and underside of the leaves are covered with silvery hairs. (*June—September*).

Least Cinquefoil (*Sibbaldia procumbens*). This small Cinquefoil, only 2 to 4 inches tall, is very rare, being found most often in the Scottish highlands. It has rather grey leaves divided into 3 leaflets, and small flowers, sometimes without petals. (*July—August*).

5 Creeping Cinquefoil (*Potentilla reptans*). The name 'Cinquefoil', meaning five-leaved, applies better to this species than some. It is a familiar garden weed, which spreads rapidly by creeping, rooting stems. Its five-petalled flowers are twice as large as those of Tormentil. It is easily confused with the next species with which it may hybridize. (*June—September*).

Trailing Tormentil (*P. anglica*). A less common intermediate between Tormentil and Creeping Cinquefoil, with both of which it hybridizes. It differs from Tormentil in having short-stalked leaves, sometimes rooting stems, and usually five-petalled flowers. It differs from Cinquefoil in being more bushy and branched, and in having some leaves with less than 5 leaflets and some flowers with only 4 petals. (*June—September*).

Sulphur Cinquefoil (*P. recta*). A very erect plant, up to 2 feet tall, and rather hairy. It usually has five-fingered leaves, and flowers nearly an inch across. Although still rare, it seems to be increasing on dry, grassy, and waste places. (*June—August*).

6 Agrimony (*Agrimonia eupatoria*). Unlike most Rosaceae, Agrimony has no epicalyx (See Herb Bennet). Its fruits carry little hooks which catch on to things and serve to disperse the seeds (Fig. 1). In old days it was used to treat snake-bite, and the flowers put into lemonade were used to cure colds. It is common in grassy places, growing from 1 to 2 feet tall. (*June—August*).

Fragrant Agrimony (*A. odorata*). This less common plant is like a very stout Agrimony, but with greener, more fragrant leaves, larger, paler flowers, and more bell-shaped fruits with the outer row of hooks turned upwards (Fig. 2). It usually grows on acid soils in shady places. (*June—August*).

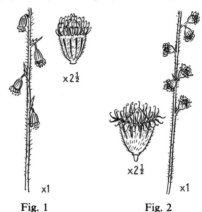

×2½ ×2½

×1 ×1

Fig. 1 Fig. 2

LIFE SIZE

1 HERB BENNET   2 SILVERWEED   3 SHRUBBY CINQUEFOIL

4 TORMENTIL   5 CREEPING CINQUEFOIL   6 AGRIMONY

PAPILIONACEAE: PEA FAMILY (see p. 211)

1 Dyer's Greenweed (*Genista tinctoria*). A plant once cultivated because a greenish-yellow dye was obtained from the leaves and stems — as both English and Latin names suggest (*tinctoria* means used in dyeing). It is now no longer used for this purpose, but still may be found wild in grassy places on clay and chalk in England and Wales and occasionally in Scotland. It has a bushy growth, about a foot high, and is rather like a small broom. (*June—August*).

Petty Whin or Needle Furze (*G. anglica*). A smaller and more wiry plant than Dyer's Greenweed, which carries spines on the stem and has fewer smaller flowers on each spike and small oval leaves. It grows on heaths and moors, but is not common. (*May—July*).

Hairy Greenweed (*G. pilosa*). This rare plant of west Wales, Ashdown Forest, and the extreme south-west of Cornwall is like a prostrate form of Dyer's Greenweed, but with small, dark green, oval leaves and smaller spikes of flowers. The pods are downy. (*May—June*).

2 Broom (*Sarothamnus scoparius*). One of the distinctive features of this shrub is its angular green stems which for much of the year do the work of leaves. Tiny leaves, sometimes three-lobed, are produced on the new shoots each year, but they often drop very early. Broom has been used for broom-making for many centuries; *Sarothamnus* means a shrub growing like a broom and *scoparius* means a broom made of twigs. It was also used medicinally, and was believed to have the power of keeping witches away. The golden light in its petals is caused mainly by the reflection of the orange-red stamens. It is found throughout the British Isles on lime-free soils. Occasionally on southern coasts a prostrate, though still shrubby, type is found. (*May—June.*)

Spanish Broom (*Spartium junceum*). A taller, stouter broom, with round stems and simple leaves, which is often planted by roadsides and railways. It flowers later than the Common Broom, and its pods are hairy all over instead of only on the edges. (*July onwards*).

3 Gorse (*Ulex europaeus*). Gorse is also called Furze or Whin. Like Broom this shrub has green stems, but in this case the leaves are modified to form the familiar spines, except on young plants which bear tiny, three-lobed leaves. It is sometimes used as a winter feed for cattle, the spines having been thoroughly crushed. It was also at one time used as a fuel. The ripe pods burst open with a loud pop, throwing the seeds several feet. On a hot summer day an almost continuous sound of popping can be heard in a gorse patch. It grows up to 8 feet high in large masses in rough grassy places and heaths throughout the British Isles, preferring lighter lime-free soils. Flowers may be found at all times of the year — hence the saying 'kissing's out of season when gorse is out of bloom'; but it flowers best *April—June*.

4 Dwarf Furze or Petty Whin (*U. minor*). This Lesser Gorse is smaller, with sprawling stems and softer, less prickly spines. It is less common, growing on heaths and moors most frequently in the east of England. (*July—September*).

Western Gorse (*U. gallii*). A west-country gorse, also sometimes called Dwarf Furze or Petty Whin, and intermediate in size and growth between Gorse and *Ulex minor*. It is very spiny, the spines being hard and sharp-pointed. It grows only on acid soils and not usually anywhere except the west of England and Wales. *Ulex minor* and *Ulex gallii* are examples of two similar plants growing in different parts of the country but looking sufficiently alike to have the same common name. (*July—September*).

1 DYER'S GREENWEED
3 GORSE

2 BROOM
4 DWARF FURZE

LIFE SIZE

19

PAPILIONACEAE: PEA FAMILY (see p. 211)

1 **Common Birdsfoot** (*Ornithopus perpusillus*). An annual plant so named because its groups of beaked pods look rather like a bird's claw. The shape of the leaves and the size and pale colour of the flowers distinguish Common Birdsfoot from Birdsfoot-trefoil, which has similar groups of pods. It is widespread and common in dry places, especially on sand and gravel, except in the extreme north and parts of Ireland. (*May—August*).

2 **Birdsfoot-trefoil** (*Lotus corniculatus*). This very gay little flower has more than seventy common names, such as Bacon-and-Eggs, Fingers-and-Thumbs, Lady's Fingers, Lady's Slipper, and Tom Thumb. The shape of both leaves and pods (Fig. 1) distinguish it from Horse-shoe Vetch (p. 22). It is very common in fairly dry grassy places. (*May—September*).

A less common, more slender, and often taller species, Slender Birds-foot-trefoil (*L. tenuis*), is found on heavier soils, especially near the sea. The stems are more branched, and the flowers are smaller, narrower, and seldom more than 4 in each head. (*June—August*).

x1   Fig. 1

Large Birdsfoot-trefoil (*L. uliginosus*). This plant is larger and more luxuriant than Birdsfoot-trefoil, and has darker green and broader leaflets. It has hollow stems and produces runners. It is usually very hairy and has more flowers in each head. It grows in damper places and is a little less common. (*June—August*).

3 **Least Birdsfoot-trefoil** (*L. angustissimus*). A rare plant found only south of the Thames, in dry, grassy places. There are only one or two flowers in each head and the straight, slender pods are about an inch long. (*May—August*).
Another uncommon species of dry grassland is Hairy Birdsfoot-trefoil (*L. hispidus*), which differs from the last species in having deep orange flowers, and pods only ½ inch long. It is found near the sea in the south-west. (*June—September*).

4 **Hop Trefoil** or Hop Clover (*Trifolium campestre*). The yellow-flowered Trefoils can be distinguished from the Medicks, which they much resemble, because the dead flowers of the Trefoils do not drop but hang covering the fruit, while those of all Medicks do drop. The many small pale-yellow flowers of Hop Trefoil turn pale brown when they die making the flower-heads look

rather tiny like hop cones. Hop Trefoil is a common plant of grassy places and waste ground. (*May—September*).

Lesser Yellow Trefoil (*T. dubium*). This plant, perhaps the true Irish Shamrock, differs from Hop Trefoil in being smaller, with looser heads of flowers and fewer flowers per head. The leaves and stem sometimes turn purplish. It is common in dry grassy places. (*May—October*).
An even smaller and more slender plant is Slender Trefoil (*T. micranthum*), a fairly common plant in grassy places on sandy or gravelly soils in the south. It has heads of only 2 to 6 dark yellow flowers on long slender stalks.

5 **Black Medick** (*Medicago lupulina*). The leaflets of this annual plant each end in a tiny point, and this and the fruits, which become black and twisted when ripe, distinguish Black Medick from the Trefoils. It is very common in bare and grassy places and, being a good fodder crop, is frequently sown in pastures, mixed with grasses. (*April—August*).

6 **Spotted Medick** or Calvary Clover (*M. arabica*). An annual easily recognised by the spots on the leaves, though sometimes plants without spots are found. The leaflets are much bigger than those of other Medicks, and the flowers a deeper yellow. The pods are short-stalked, rounded, spiny, and twisted. It grows best near the sea but may also be found farther inland in sandy waste places. (*April—September*).

7 **Hairy Medick** (*M. hispida*). This less common plant, also found near the sea, is smaller and more hairy than Spotted Medick and has no dark blotch on its leaves. Its pods are in a flattened spiral with strongly marked lines. It is found only in southern and eastern England. (*May—September*).
A similar species is Small Medick (*M. minima*), whose spirally twisted pods are surrounded by hooked spines. It can be identified by its stipules which are untoothed. It is also scarce and grows in much the same places as Hairy Medick, though not usually flowering after July.

8 **Yellow Vetchling** (*Lathyrus aphaca*). The tendrils of this annual plant are, in fact, modified leaves, and what look like leaves are really large stipules at the base of the leaf stalk. The flowers grow on long stems, usually singly. It is a scarce plant of dry bushy and grassy places in the south. Its seeds germinate late in the year and so most seedlings are probably killed by frost each winter. (*June—August*).

1 COMMON BIRDSFOOT   2 BIRDSFOOT-TREFOIL   3 LEAST BIRDSFOOT-TREFOIL

4 HOP TREFOIL   5 BLACK MEDICK   6 SPOTTED MEDICK

7 HAIRY MEDICK   8 YELLOW VETCHLING

LIFE SIZE

PAPILIONACEAE: PEA FAMILY (See p. 211)

1 Kidney-vetch or Ladies' Fingers (*Anthyllis vulneraria*). The colour of the Kidney-vetch flowers, especially when growing near the sea, may vary from pale yellow to orange or fiery red. The plant has tiny seed-pods, each carrying only one seed and enclosed in a calyx covered with woolly hairs (Fig. 1). Kidney-vetch is common in dry, grassy places, especially on chalk and limestone and near the sea. (*May—August*).

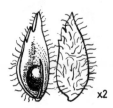

Fig. 1.

2 Horse-shoe Vetch (*Hippocrepis comosa*). This plant can be distinguished from Birdsfoot-trefoil (p. 21) by its leaves, which are narrower and have more numerous leaflets, and by its flowers, which are smaller. Its name 'horse-shoe' refers to the pods which are long and divided into distinct horse-shoe-shaped segments, each segment containing one seed. These separate when ripe. The plant grows in chalk and limestone turf and is common in suitable localities, though not found in Ireland. (*May—July*).

3 Milk Vetch or Liquorice (*Astragalus glycyphyllos*). This plant is not the liquorice used in making certain sweets and laxatives. Its heads of large dirty-looking flowers on short stalks distinguish it from other Vetches and Peas. The stout, curved pods may reach 1½ inches in length. It grows up to 3 feet high in rough grass in sunny places on chalk and limestone. It is not common and never found in Ireland. (*June—August*).
There are two other species of this genus, the Purple Milk Vetch (*A. danicus*) and Alpine Milk Vetch (*A. alpinus*), which are both rare. The former bears rich violet flowers erect on a prostrate plant and grows on chalk and limestone from the Chilterns to Yorkshire and N.E. Scotland. The latter is a smaller, slenderer plant with paler flowers, and grows only on a few mountain ledges in the E. Highlands.

4 Melilot (*Melilotus officinalis*). The Latin name, *officinalis*, means 'used medicinally'. Melilot was prob-ably introduced to Britain for use in making poultices. It gives off a lovely smell of new-mown hay as it dries. The ripe pods are brown and wrinkled (Fig. 2a). It grows in fields and waste places in southern England and Ireland. It is often quite tall, up to 4 feet in height. (*June—September*).

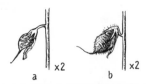

Fig. 2

Golden Melilot (*M. altissima*). A very similar plant to Melilot but with deeper yellow flowers and more compact flower-heads. The larger ripe pods are black and hairy (Fig. 2b). It usually grows in woods and waste places on heavier soils. (*June—August*).

Small-flowered Melilot (*M. indica*). This plant looks much like a very small Common Melilot — not more than a foot high, and the flowers are tiny and pale yellow. The pods are olive green when ripe. It grows in the same kind of places as Common Melilot but is less common. (*June—September*).

White Melilot (*M. alba*). A plant first introduced as a forage plant under the name of Bokhara Clover. It is now spreading by roadsides and on waste ground, mainly in southern England and Wales. It looks like Common Melilot but has smaller, white flowers, longer leaflets, and grows taller. (*June—August*).

5 Meadow Vetchling or Meadow Pea (*Lathyrus pratensis*). Although this Pea has thin, rather weak stems, it may reach up to 3 feet in height by scrambling over other plants, climbing by means of its tendrils. These tendrils are, in fact, modified leaves. There are also green leafy stipules at the base of the leaf stalk. The pods are rather flattened when ripe and contain 5 to 10 seeds. (*June—August*).

1 KIDNEY-VETCH

2 HORSE-SHOE VETCH

3 MILK VETCH

4 MELILOT

5 MEADOW VETCHLING

LIFE SIZE

SCROPHULARIACEAE: SNAPDRAGON FAMILY (See p. 216)

1 Common Cow-wheat (*Melampyrum pratense*). All the Cow-wheats are semi-parasites: that is, they grow attached to the roots of other plants, obtaining part of their nourishment from them. Common Cow-wheat has the distinguishing characteristic that all its flowers turn to face the same way. In colour they range from deep yellow to almost white and are sometimes tinged with red or purple. The plant is from 6 to 12 inches tall and is common in woods and on moors. (*May—September*).

Small Cow-wheat (*M. sylvaticum*). This is a more slender plant and much less common than the last species, with smaller, deep yellow flowers which are set further apart. The lower lip of the flower tube is always turned downwards. It is found in hilly northern districts. (*June—August*).

Crested Cow-wheat (*M. cristatum*). A beautiful uncommon plant found on the edges of woods in East Anglia. It has purple, fine-toothed bracts (the crests) beneath the flowers (Fig. 1), the lower ones having long trailing green points. The flowers, shaped like other Cow-wheats, are yellow and purple and face in all directions instead of all being turned one way. (*June—September*).
A similar, very rare species, Field Cow-wheat (*M. arvense*), is a striking plant with spiky-looking purple bracts and pink and yellow flowers. It was once a cornfield weed, its seeds giving a bluish colour and bitter taste to the flour; but now it has been almost eradicated and is found only occasionally in southern cornfields.

Fig. 1

2 Yellow Rattle (*Rhinanthus minor*). A very variable plant both in height (3 inches to 2 feet) and in growth. Its toothed, unstalked leaves are sometimes very narrow and sometimes nearly 1 inch broad. The stem is stiff and 4-angled. After the flowers die, the seeds ripen and are shed into an enlarged calyx, which forms the 'rattle'. Eventually the seeds are shaken out of the calyx on to the ground. It is a semi-parasite found in grassy places and woods throughout the country. (*May—July*). There is an extremely rare, later-flowering species, *R. major*, which is taller, bushier, and with larger, more crowded flowers.

Yellow Bartsia (*Parentucellia viscosa*). This uncommon, small plant is easily distinguished from Yellow Rattle because it has a smaller calyx which does not swell after flowering, and because its unstalked, toothed leaves

Fig. 2

alternate up the stem. It is also characterized by its sticky hairs (Fig. 2). It grows chiefly in damp, grassy places near southern and western coasts. (*June—October*).

3 Common Toadflax (*Linaria vulgaris*). This Snapdragon-like plant can become a pest in gardens, difficult to get rid of as tiny pieces of root can grow into new plants. It differs from the garden Snapdragon, *Antirrhinum majus*, in having long spurs and smaller flowers. *A. majus*, also, is usually red when wild. Toadflax is common in grassy and waste places, but is less frequent in Ireland. (*July—October*).
There is also a very rare prostrate species of Toadflax, *L. supina*, which is smaller than common Toadflax and is found mainly in the south-west.

4 Round-leaved Fluellen (*Kickxia spuria*). A rather uncommon plant, found only in southern England and Wales on light soils, principally chalk. Like Ivy-leaved Toadflax (p. 141) the flowers resemble tiny Snapdragons with curved spurs. The seeds escape from the ripened capsule through pores with little lids which drop off. (*July—October*).

Sharp-leaved Fluellen (*K. elatine*). This plant, which spreads further northwards than Round-leaved Fluellen, is less hairy and has smaller, arrow-shaped leaves and smaller, paler flowers with straight spurs (Fig. 3). (*July—October*).

Fig. 3

5 Monkey-flower (*Mimulus guttatus*). This beautiful plant of wet places, especially stream-banks, was originally a native of the Aleutian Islands, and was first grown in gardens in the early 19th century. It soon began to naturalize in Wales and has since spread throughout the British Isles. It grows up to 18 inches tall, with stalked lower leaves and unstalked upper leaves clasping the stem. Some people fancy that its flowers resemble little monkey faces. (*June—September*).
There are two other more uncommon, closely related species, both smaller than Monkey-flower. The first, *M. luteus*, has longer, hairless flower stalks, and the flowers are marked with large red or purple blotches, sometimes appearing almost entirely purple — hence its English name Blood-drop Emlets. It is found mainly in Scotland. The second, *M. moschatus*, is a pale plant covered with sticky hairs, and its flowers are unspotted. It is a native of North America and was planted in English gardens for the sake of its musky smell — hence its English name, Musk. It naturalized in a few places but inexplicably has lost its smell.

LIFE SIZE

1 Common Cow-wheat   2 Yellow Rattle   3 Common Toadflax

4 Round-leaved Fluellen   5 Monkey-flower

PRIMULACEAE: PRIMROSE FAMILY (See p. 215)

1 Yellow Loosestrife (*Lysimachia vulgaris*). This is a large and upstanding plant, sometimes growing as much as 4 feet high, and the picture shows only the upper part. The leaves grow in pairs, threes, or fours, and are often dotted with black glands. The red stamens are opposite the petals, not the sepals, a characteristic feature of this family. It grows in wet places, especially by rivers and ponds, and is most common in the south. It is not related to Purple Loosestrife (p. 123), although the two may grow together. (*July—August*).

Tufted Loosestrife (*L. thyrsiflora*). A plant of fens and lakes in the north and very rare elsewhere. The flower stems grow from 1 to 2 feet high and bear pairs of long, unstalked leaves, at the base of which spring short stalks carrying bunches of small, feathery flowers (Fig. 1). The stamens are prominent and a brighter yellow than the flowers. (*June—July*).

Fig. 1

2 Creeping Jenny (*L. nummularia*). *Nummularia* means 'coin-like', for the leaves are nearly round. This is probably the reason for its other English names, Herb Tuppence and Pennywort, the latter name being also used for two other round-leaved plants (pp. 47 & 53). Creeping Jenny is common in damp woods and grassy places, especially in the south. It must spread by sending out runners, as it has never been known to produce seeds in Britain. (*June—August*).

3 Yellow Pimpernel (*L. nemorum*). This differs from Creeping Jenny in having more pointed leaves and smaller, deeper yellow flowers on longer, more slender stalks. This plant also grows in damp places but can tolerate drier soil and hotter sun than can Creeping Jenny. It also creeps, but does produce seeds as well. (*May—September*).

4 Primrose (*Primula vulgaris*). A well-known spring flower often grown in gardens, not only in the pale yellow form but also in pink, purple, or white flowered varieties. The lines of darker yellow pointing to the centre of the flower are called honey guides, and are supposed to guide bees to the tube, at the base of which nectar is produced. Primroses, Cowslips, and Oxlips produce two kinds of flower, pin-eyed and

thrum-eyed. In the former (Fig. 2a) the pistil is long-stalked with the stigma visible at the mouth of the tube, and the stamens are attached out of sight lower down the tube. In the thrum-eyed flowers (Fig. 2b) the pistil is short-stalked and hidden by the stamens which are attached at the top of the tube. This makes self-pollination difficult, but the fact that the pistil of the one type of flower is on a level with the stamens of the other type means that they are touched by the same part of the pollinating bee, and the pollen transferred from the one type to the other, causing cross-pollination. Primroses grow throughout the British Isles, especially in the west, in woods and hedgebanks and on railway banks and similar grassy places. They are less common than they once were near large towns because so many have been picked and dug up. (*March—May.*)

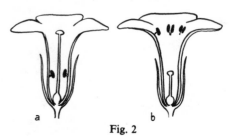

a   b

Fig. 2

5 Cowslip (*P. veris*). In many parts of the country the Cowslip is called Paigle because the heads of nodding flowers are thought to symbolize St. Peter's bunch of keys. The leaves grow in a rosette like those of the Primrose but are rather smaller. It is widespread in the south in meadows and pastures, especially on chalk, limestone, or clay, but less frequent in the north. (*April—May*).

6 Oxlip (*P. elatior*). A flower of damp woods on chalky clay in East Anglia, where it is often more common than the Primrose. It is hardly found anywhere else. It can be easily confused with the False Oxlip, but the flowers of the true Oxlip tend to hang down from one side of the stalk rather like those of the Cowslip. It has a long stem, up to 12 inches, tapering leaves, and a habit of growing in masses instead of as scattered plants. (*April—May*).

7 False Oxlip (*P. veris x vulgaris*). This is not a separate species but a hybrid between the Cowslip and Primrose, and wherever both grow a few False Oxlips may be found. It is more hairy than the Cowslip, with larger, paler flowers and leaves which taper more gradually towards the base. (*April—May*).

LIFE SIZE

1 YELLOW LOOSESTRIFE   2 CREEPING JENNY   3 YELLOW PIMPERNEL
4 PRIMROSE   5 COWSLIP   6 OXLIP
7 FALSE OXLIP

1 Yellow Flag (*Iris pseudacorus:* family *Iridaceae*). A common plant to be found throughout the British Isles by pools and rivers and in marshy places. It may grow as much as 6 feet tall. The plant depends on cross-pollination, the parts of the flower being so constructed that self-pollination is prevented. (*May—August*).

2 Daffodil (*Narcissus pseudonarcissus:* family *Amaryllidaceae*). This is the true wild Daffodil, though garden Daffodils, Narcissi, and hybrids are sometimes found growing apparently wild. The wild Daffodil, which seldom grows more than 12 or 14 inches tall, is found locally in damp meadows, especially in the south-west. It is often called the Lent Lily because of its time of flowering. (*March—April*).

3 Yellow Archangel (*Galeobdolon luteum:* family *Labiatae*). This common plant spreads by means of long leafy runners which are usually produced after the flowering period. Except for the yellow flowers, it is much like White Dead-nettle (p. 97), but is less hairy and has narrower, darker green leaves. It prefers woods and shady places on heavy soils, and is found more commonly in the south than in Scotland or Ireland. (*May—June*).

Large Hemp-nettle (*Galeopsis speciosa*). A showy plant whose large pale-yellow flowers have a striking purple blob on the lower lip, and two small lumps between the outer and central lobes — a characteristic of Hemp-nettles (Fig. 1). Unlike Yellow Archangel, the flowers are crowded at the top of the stem and the plant is more hairy. It is not common in the south but is found more often on arable land in the north. (*July—September*).
A very rare species, now found only in a part of North Wales, is Downy Hemp-nettle (*G. dubia*). This has very long, pale yellow flowers, less crowded together than those of Large Hemp-nettle.

Fig. 1

4 Great Mullein or Aaron's Rod (*Verbascum thapsus:* family *Scrophulariaceae*). All British Mulleins are normally biennials with a rosette of leaves from which the flower stalk grows in the second year. Great or Common Mullein, the largest of the Mulleins, often grows 4 feet or more tall. It is common on dry soils in sunny places, except in northern Scotland. The base of the woolly stem-leaves extends down the stem as wings. The almost stalkless flowers are nearly 1 inch across, and have 5 stamens, 3 covered in white woolly hairs and 2 almost hairless. (*June—August*).
A much less common species, Orange Mullein (*V. phlomoides*), may be confused with Great Mullein, but has a less woolly appearance, larger and often more orange flowers, and stem leaves which do not form wings down the stem.
Another uncommon species, White Mullein (*V. lychnitis*), found on chalky soils in south-east England, is smaller and more slender than Great Mullein and is covered with greyish hairs. The flowering stems are branched and without the leaf-wings, and the flowers are white, except for a yellow form found in Somerset.

Dark Mullein (*V. nigrum*). This Mullein, which grows only 6 to 18 inches high, has stalked, slightly hairy leaves, deep yellow flowers, and conspicuous stamens covered with purple hairs. It is not uncommon in the south on chalk or sand, but rare in the north. (*June—October*).

Large-flowered Mullein or Twiggy Mullein (*V. virgatum*). A rare plant found sometimes in waste places in the south. The lower parts of the plant are almost hairless, but the upper parts are covered with sticky hairs. The flowers are as large as those of Great Mullein but paler yellow, not so closely packed together, and with orange anthers and conspicuous purple hairs on the stamens. (*June—August*).
A smaller and less sticky species, Moth Mullein (*V. blattaria*), is equally uncommon. It has looser spikes of flowers which may be yellow but are often pale pink or white.

5 Ground-pine (*Ajuga chamaepitys:* family *Labiatae*). This small hairy plant both looks rather like a Pine seedling, and has a piney smell. The stem is sometimes purple, and the flowers have tiny red spots on their lower lips. It is a rare plant of chalk soils in south-east England. (*May—September*).

6 Henbane (*Hyoscyamus niger:* family *Solanaceae*). This very poisonous and coarse-looking plant has a foul smell. It grows up to 4 feet tall and is covered with whitish, sticky hairs. All parts of the plant contain a narcotic drug called hyoscine, which is still sometimes used medicinally, but today has been largely replaced by morphine and codeine, obtained from the Opium Poppy. Henbane grows not very commonly in bare or disturbed ground, often in farmyards on sandy soil, chiefly in the south near the sea. (*June—August*).

TWO-THIRDS LIFE SIZE

1 YELLOW FLAG   2 DAFFODIL   3 YELLOW ARCHANGEL

4 GREAT MULLEIN   5 GROUND-PINE   6 HENBANE

1 **Crosswort** (*Galium cruciata*: family *Rubiaceae*). This is a softly hairy plant all over and easily distinguished from Yellow Bedstraw because its growth is more erect, the flower stalks being 9 to 15 inches tall, and because it has broader leaves borne in groups of four, forming a cross. The tiny four-petalled flowers are sweet-scented. It grows throughout the British Isles in bushy or grassy places, especially on chalk or limestone, but it is rare in Ireland. (*April—June*).

2 **Yellow Bedstraw** or Lady's Bedstraw (*G. verum*). A sprawling plant, smelling of hay, and with honey-scented flowers. In old days, people often used to sleep on a mattress of dried bedstraw and other plants, covered with a sheet. Such a mattress smelt pleasant and could be easily burnt and renewed when it became soiled. Bedstraw flowers were also used for curdling milk to make cheese, and a red dye was extracted from the stems. It is very common in grassy places throughout the British Isles. (*July—August*).

3 **Yellow-wort** (*Blackstonia perfoliata*: family *Gentianaceae*). Unlike the other members of this family (see pp. 125, 139 and 177), Yellow-wort has up to 8 sepals, petals, and stamens, instead of the more usual 4 or 5. The stem leaves grow in pairs joined together round the stem. The fruit is an oval capsule. It is fairly common in southern England in chalk and limestone turf, and rather less common elsewhere. (*June—October*).

4 **Yellow Corydalis** or Yellow Fumitory (*Corydalis lutea*: family *Fumariaceae*). This plant is a native of Southern Europe; it has been grown in gardens for so long that it has escaped and become naturalized, but it still does not spread far from cultivated places. It is especially frequent on walls. The flower stalks, although they are attached all round the stem, twist so that the flowers all face one side. (*July—September*).

5 **Wood Sage** (*Teucrium scorodonia*: family *Labiatae*). Unlike most of the rest of its family (see pp. 97, 143 — 149 and 177), Wood Sage has flowers without an upper lip so that the stamens are very prominent. It once had a number of medicinal uses. A tea made from the dried leaves was drunk, and maybe still is in Gloucestershire, as a preventive against rheumatism. The plant, which grows about 1 foot high, is common throughout Great Britain in woods, grassland, heaths, and dunes, preferring dry, shady places on rather acid soils. (*July—September*).

6 **Greater Bladderwort** (*Utricularia vulgaris*: family *Lentibulariaceae*). A water plant so called because it has little bladders on its leaves. These have two interesting functions. They are filled with air and so keep the plant floating near the surface of the water. They also trap tiny water insects, which in time die and decay and produce mineral salts, which the plant absorbs as food. Greater Bladderwort is not a common plant but is found throughout the British Isles in lakes, ponds, and ditches, usually in relatively deep water. It rarely flowers and then only in the south. It passes the winter in the form of small buds, which drop to the bottom of the pond and grow into new plants in the spring. In this way it spreads and increases without producing seeds. (*July—August*).

Irish Bladderwort (*U. intermedia*). This is a rare Bladderwort belonging almost entirely to the bog pools of North Scotland and Ireland. It grows in shallower water than Greater Bladderwort and is rather smaller. It rarely flowers. It has two kinds of leaves — green feathery leaves with toothed, bristly segments but with very few bladders, and colourless leaves bearing many bladders and often buried in the mud (Fig. 1). (*July—September*).

x1

Fig. 1

Lesser Bladderwort (*U. minor*). This smallest of the Bladderworts also has two kinds of leaves, like Irish Bladderwort, but the green leaves are feathery without teeth and bristles (Fig. 2). The flowers are much smaller and are pale yellow with a short spur. It grows in shallow water, especially in bog pools, throughout the British Isles, but is rare and local. (*June—September*).

x1

Fig. 2

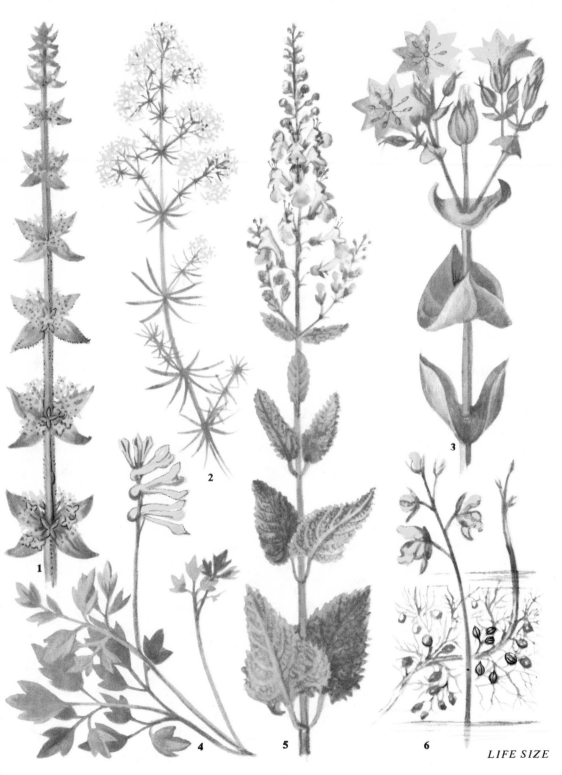

LIFE SIZE

1 CROSSWORT   2 YELLOW BEDSTRAW   3 YELLOW-WORT

4 YELLOW CORYDALIS   5 WOOD SAGE   6 GREATER BLADDERWORT

COMPOSITAE: DAISY FAMILY (See p. 218)

1 **Nodding Bur-marigold** (*Bidens cernuus*). This annual, so called because the round drooping flower-heads tend to nod, grows up to 2 feet high and is fairly common near ponds and streams and in wet places. Most conspicuous are the long leaf-like bracts below the flower-heads. The fruits, which are achenes like those of all Compositae, carry three or four barbed bristles which serve to disperse the seeds by catching on to clothing, or the coats of animals (Fig. 1).

×2

Fig. 1

Another common Bur-Marigold (*B. tripartitus*) has short-stalked leaves which, as the Latin name indicates, are mostly deeply divided into three lobes. The flower-heads are smaller and held more erect. (*July—September*).

2 **Golden Rod** (*Solidago virgaurea*). A common plant, growing up to 2½ feet high, found in dry woods and on dunes, heaths, and hedge-banks throughout the British Isles, though rarely in the south-east. (*July—September*).
The garden Golden Rod (*S. canadensis*) often escapes and grows wild. It is much taller and more hairy, and the mass of flower-heads grows in the shape of a pyramid. (*August—October*).

3 **Wormwood** (*Artemisia absinthium*). The grooved stems and grey leaves are covered with silky hairs. The lowest leaves are very much subdivided and the upper stem leaves much less or not at all. Wormwood is used to make various liqueurs, such as absinthe. Like Mugwort and Sea Wormwood (See p. 60), it has an aromatic scent and used to be used to keep away moths and other insects. It grows throughout Europe and Asia and is found infrequently in waste places in the British Isles. (*July—September*).

4 **Wayside Cudweed** or Marsh Cudweed (*Gnaphal um uliginosum*). A hairy, much-branched annual which grows from about 2 to 8 inches high with creeping or upright stems. The leaves are woolly on both sides. The flower-heads, in clusters of three to ten, are small, with brown bracts and yellowish florets. It is widespread and common on damp places in sandy fields, heaths, and waysides. (*July—September*.)

A very rare species of the Channel Islands and a few places in East Anglia and Hampshire is Jersey Cudweed (*G. luteo-album*). This has no leaves at the base of the flower clusters, pale yellow bracts, and yellowish florets with red stigmas.

Wood Cudweed or Heath Cudweed (*G. sylvaticum*). The flower-heads are larger than in the last species and are arranged in a long spike instead of in clusters at the end of the stem (Fig. 2). The leaves are smooth above and hairy beneath. The florets are yellowish-brown with brown and green striped bracts. The plant is fairly common in open woods and heaths. (*July—September*).
There is a rare species (*G. norvegicum*) which is shorter and woollier, with more compact flower-spikes. It is found only in the Scottish Highlands. Another rare Highland species is Dwarf Cudweed (*G. supinum*), a small tufted plant, about 2 inches high, with one to three flowering heads at the end of the stems.

×½

Fig. 2

5 **Slender Cudweed** or Field Cudweed (*Filago minima*). An inconspicuous annual with stem and leaves covered with grey silky hairs. The flower-heads are in clusters of three to six and have yellowish-tipped bracts which spread out like a star when the plant is fruiting. Slender Cudweed is locally common on heaths and dry sandy fields, but is rare in the far north. (*June—September*).

6 **Common Cudweed** or Upright Cudweed (*F. germanica*). This woolly-looking annual plant is thickly covered with white hairs which nearly hide the clusters of flower-heads. The leaves usually have wavy edges. Stems branch from just below the flower-heads. The inner bracts round the flower-heads have yellow tips. This Cudweed is common in dry sandy places, becoming rarer in the north. (*July—September*).
There are two other closely related species which are found occasionally in Southern England. Red-tipped Cudweed (*F. apiculata*) is recognised most easily by the red tips to the bracts round the flower-heads. It also has broader leaves and yellowish hairs. Broad-leaved Cudweed (*F. spathulata*) is recognised by its stems branching near the base of the plant and growing along the ground. The leaves are spatula-shaped, as the Latin name suggests.

LIFE SIZE

1 Nodding Bur-marigold   2 Golden Rod   3 Wormwood

4 Wayside Cudweed   5 Slender Cudweed   6 Common Cudweed

COMPOSITAE: DAISY FAMILY (See p. 218)

1 **Bristly Ox-tongue** (*Picris echioides*). A leafy annual or biennial, up to 3 feet tall, and containing milky juice. It is covered with stiff prickly hairs which, under a lens, can be seen to have hooks at their ends. There are 3 to 5 leafy outer bracts below the flower-heads, resembling a calyx, and about 10 narrower, pointed, inner bracts. The fruits are beaked and have a pappus of white feathery hairs. This locally common plant frequents roadsides, fields, and waste places on heavy or chalky soil, though not in northern Scotland. (*June—October*).

Hawkweed Ox-tongue (*P. hieracioides*). Similar to Bristly Ox-tongue, but with short, narrow, outer bracts round the flower-heads (Fig. 1), and only slightly beaked fruits. The pappus is cream-coloured. It is locally common by roadsides and on waste ground in England, Wales, and south Scotland, mainly on chalky soil.

×1

Fig. 1

2 **Corn Sowthistle** or Field Milk-thistle (*Sonchus arvensis*). A perennial with a hollow stem and milky juice, growing about 3 feet high. The lower leaves are oblong and lobed with prickly teeth. The stem leaves have rounded lobes at the base. The flower stalks and bracts below the large flower-heads are covered with yellow-tipped hairs. The fruits are unbeaked, and the pappus consists of 2 equal rows of simple hairs. This Sowthistle is common in cornfields, marshes, and on waste ground throughout the British Isles. (*July—September*).

Marsh Sowthistle (*S. palustris*). This very tall, very rare perennial of south-east England grows to 3 feet or more in height. It can be distinguished from Corn Sowthistle by its thick stems and its stem leaves, which have short pointed lobes at the base and tiny teeth. The hairs on the flower stalks and involucre are usually blackish-tipped. (*July—September*).

3 **Common Sowthistle** or Smooth Sowthistle (*S. oleraceus*). An annual with deeply-lobed, toothed, lower leaves and long, pointed lobes at the base of the upper leaves. Neither leaves nor stems are prickly. The outer florets of the small flower-heads are often tinged with purple underneath. This is a very common weed of cultivated land throughout the British Isles. (*May—October*).

Prickly Sowthistle (*S. asper*). A plant often found with Common Sowthistle, though it is not quite so common. It can be recognized by the unlobed or only slightly lobed lower leaves, the rounded ear-shaped lobes at the base of the stem leaves, and the prickly teeth round the leaf edges. (*June—October*).

4 **Tansy** (*Tanacetum vulgare*). This perennial, 1 to 3 feet high, has stiff, ribbed stems, leaves covered with small dots (glands), and a lemon-like scent. The numerous flower-heads often form a flat-topped mass. Each head has 3 or 4 rows of bracts and many small florets. Tansy is fairly common on roadsides and waste places throughout the British Isles. It was once grown in cottage gardens for use in cooking and as a tonic. (*July—September*).

5 **Ploughman's Spikenard** (*Inula conyza*). A softly hairy biennial, growing up to 3 feet high, with very slightly toothed alternate leaves. The lower stem leaves are stalked, but the upper ones are tapered to the base and unstalked. The inner row of bracts round the flower-heads are reddish. This fairly common plant is found in open woods and hedgerows and on dry, stony places on chalky soil in England and Wales. (*July—October*).

1 BRISTLY OX-TONGUE   2 CORN SOWTHISTLE   3 COMMON SOWTHISTLE
4 TANSY   5 PLOUGHMAN'S SPIKENARD

LIFE SIZE

These yellow Dandelion-like flowers of the Compositae family all have typical flower-heads composed of a group of tiny flowers massed together. Surrounding each composite head in place of sepals are several rows of green bracts, forming what is called an 'involucre'. In order to name these plants correctly it is often necessary to use a lens to see small features such as hairs and scales. This simplified diagram of an imaginary section through a flower shows the important parts to look for in identifying a species (Fig. 1).

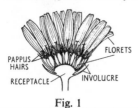

Fig. 1

Hawk's-beards (*Crepis*). The branched flowering stems bear clasping leaves. The pappus hairs are unbranched, usually white and silky (Fig. 2a). They differ from Hawkweeds in the narrowed or beaked upper part of the fruits and in the one or two rows of sepal-like bracts. The two most common species are described here.

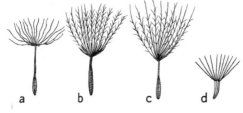

Fig. 2

1 **Beaked Hawk's-beard** (*C. taraxacifolia*). A tall plant, up to 2½ feet. The numerous flower-heads each have one row of hairy bracts, with a few spreading bracts at the base of the involucre. The ripe fruits (achenes) have long, slender beaks. This plant is locally common in waste places in England and Wales, especially on chalk and limestone. (*May—July*).
2 **Smooth Hawk's-beard** (*C. capillaris*). This differs from Beaked Hawk's-beard in having few flower-heads, arrow-shaped stem leaves, and achenes without beaks. The small forms found on heaths often have no stem leaves. It is common in fields and waste places. (*June—November*).

Hawkbits (*Leontodon*). This group has leafless stems. The pappus is usually composed of two rows of hairs, the inner ones feathery and the outer ones unbranched (Fig. 2b). There are no scales between the florets.
3 **Autumnal Hawkbit** (*L. autumnalis*). Less than half the size of the Common Cat's Ear, this species has narrower and more shiny leaves with longer teeth. The fruits are not beaked (Fig. 2b). It is common in grassy places. (*July—October*).

4 **Rough Hawkbit** (*L. hispidus*). This is identified by the forked hairs which cover the leaves. The flowers are large, and when the fruit is ripe, the pappuses form a 'clock' like that of a small Dandelion. It is common in grassy places, especially on chalk and limestone. (*June—September*).
Lesser Hawkbit (*L. leysseri*). This smaller and less hairy species is found especially on commons and dunes. The tuft of leaves are often reddish and the bracts forming the involucre are brown edged. (*June—October*).

Cat's Ears (*Hypochaeris*). This group has leafless and hairless stems. There are narrow scales between the florets, and a pappus of feathery hairs (Fig. 2c).
5 **Common Cat's Ear** (*H. radicata*). This species, about 1 foot high, is covered with rough hairs and has short bracts on the stem. It is common in grassy places and can be a troublesome weed on lawns because its flat rosettes of leaves smother the grass beneath them. The fruits are mostly beaked (Fig. 2c). (*June—September*).
Smooth Cat's Ear (*H. glabra*). This is less common, mostly on sandy soils. It has shiny, reddish, almost hairless leaves and smaller, less conspicuous flower-heads, which open only in bright sunshine.
Spotted Cat's Ear (*H. maculata*). An also uncommon, rather taller Cat's Ear which grows in chalk and limestone turf in eastern England and on western sea cliffs. It has broader leaves, often carrying purplish-black spots. The flower stem carries one large head with a hairy involucre.

Hawkweeds (*Hieracium*). The stems usually carry leaves of variable shapes. The pappus hairs are unbranched and usually brownish (Fig. 2d). There are many varieties which are difficult to distinguish. We describe three groups here, and a fourth, Orange Hawkweed, on page 104.
6 **Leafy Hawkweed** (*Aphyllopoda* group). Late-flowering Hawkweeds with leaves on the stems and no basal rosette of leaves belong to this group. They are tall, up to 4 feet, and usually bear a number of flower-heads on branched stems. Common, especially in sandy and heathy places and on riverbanks. (*July onwards*).
7 **Mouse-ear Hawkweed** (*Pilosella* group). This group varies much in size, though seldom much taller than 12–15 inches. It has long, leafy runners growing out from the basal leaf rosette. The leaves are pale green, covered with long white hairs. The flowers, which grow singly on long, leafless stalks, are pale lemon yellow, often with some red beneath. Common in short turf. (*May—August*).
8 **Few-leaved Hawkweed** (*Phyllopoda* group). Earlier-flowering shorter Hawkweeds, with a basal rosette of leaves but few or none on the stem, belong to this group. The stems are usually unbranched and carry fewer flower-heads. Common on walls, dry banks, and mountain ledges. (*May—July*).

1 Beaked Hawk's-beard   2 Smooth Hawk's-beard   3 Autumnal Hawkbit
4 Rough Hawkbit   5 Common Cat's Ear   6 Leafy Hawkweed
7 Mouse-ear Hawkweed   8 Few-leaved Hawkweed

LIFE SIZE

COMPOSITAE: DAISY FAMILY (See p. 218)

1 **Corn Marigold** (*Chrysanthemum segetum*). This has handsome, easily recognized flower-heads, carried on long stalks. The lower leaves are stalked and have toothed edges or are cut into leaflets; the upper ones clasp the stem and are sometimes little toothed. The plant, which grows 6 to 18 inches tall, is an annual, locally common in cornfields throughout the British Isles. (*July—September*).

Corn Marigold may be confused with the much rarer Yellow Chamomile (*Anthemis tinctoria*) found occasionally in hot, dry places. Yellow Chamomile, however, is rather smaller and has smaller, very deeply divided leaves.

2 **Golden Samphire** (*Inula crithmoides*). A perennial, about 1 to 2 feet high, with smooth, fleshy stems and leaves. The fruits carry a pappus of white hairs. This is a rare plant of sea cliffs and rocks and salt marshes, mainly in the south and west of Britain and in south-east Ireland. (*July—September*).

Elecampane (*I. helenium*). Another rare plant, covered with short, soft hairs and bearing yellow Daisy-like flowers, which are much larger than those of Golden Samphire and have long straggly petals. The lowest leaves are about a foot long, oval and toothed, with long stalks; the upper ones are smaller and unstalked. It was originally cultivated in Britain for use in tonics and other medicines, and gradually naturalized, growing mainly in copses and damp places. (*July—August*).

3 **Common Fleabane** (*Pulicaria dysenterica*). A very woolly perennial, 1 to 2 feet high, with soft leaves clasping the stiff stems (Fig. 1*a*). There are numerous florets in each flower-head, the outer ones (ray florets) being about

twice as long as the inner ones (disk florets). The fruits carry a pappus. The plant is common in damp meadows and other wet places throughout most of Britain and Europe, extending as far as Russia and North Africa. (*July—September*).

Small Fleabane (*P. vulgaris*). A smaller plant than Common Fleabane and less thickly covered with hairs. The leaves have wavy edges, but lack the rounded basal lobes of Common Fleabane (Fig. 1*b*). The small flower-heads have almost upright outer florets, hardly longer than the inner ones (Fig. 2). This is a rare plant of damp, sandy places and edges of ponds in southern England and Wales.

Fig. 2

4 **Rayless Mayweed** or Pineapple Weed (*Matricaria matricarioides*). A small annual, usually less than 9 inches high, with a strong aromatic scent. The rounded flower-head, which might be thought of as resembling a pineapple in shape, contains a cone-shaped hollow; this can be seen if the head is cut in half downwards through the centre (Fig. 3). There are no spreading outer or ray florets, only small disk florets like those in the centre of a Daisy (*see* p. 99). This common plant grows in waste places, especially well-trodden ones such as paths, cart-tracks, and farm-yards. (*June—September*).

Fig. 3

5 **Carline Thistle** (*Carlina vulgaris*). The rosette of leaves formed by this biennial in its first year dies away in the second year before flowering time. The purplish flowering stem grows up to 12 inches high and carries 1 to 5 flower-heads. Stiff, shiny, inner bracts spread out round the flower-head like ray florets, closing up in bad weather and at night. The fruits carry a long silky pappus of feathery hairs. This plant is locally common on chalky soil, though less common in the north. (*July—September*).

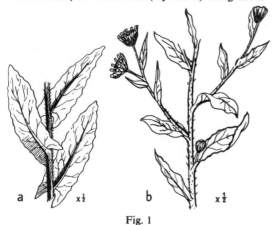

a   x⅓   b   x½

Fig. 1

LIFE SIZE

1 Corn Marigold   2 Golden Samphire   3 Common Fleabane
4 Rayless Mayweed   5 Carline Thistle

COMPOSITAE: DAISY FAMILY (See p. 218)

1 Dandelion (*Taraxacum officinale*). The very showy flower-heads are carried singly on unbranched, leafless, hollow stems, which contain a sticky white juice—as do all parts of the plant. The narrow outer bracts of the involucre spread out or bend downwards. Dandelions have long tap roots which are difficult to dig out from lawns and flower-beds (Fig. 1), where they are tiresome weeds. The ripe fruits with spreading pappus form the well-known 'clock'. (*March—October*).

There are three other types of British Dandelions: the Lesser Dandelion (*T. laevigatum*), which grows in dry places and has purplish-red fruits with rough tips; the Narrow-leaved Marsh Dandelion (*T. paludosum*), which has narrow, almost unlobed leaves and broad outer bracts pressed closely to the involucre; and the Broad-leaved Marsh Dandelion (*T. spectabile*), which has broader, hairy leaves with red midribs. Both the last two are locally common in wet places.

Fig. 1

2 Nipplewort (*Lapsana communis*). The numerous flower-heads each have an involucre of about 8 bracts, with a few more bracts at the base. The curved achenes have no pappus. The flower-heads, which are visited by insects on sunny mornings, close in the afternoons and in dull weather, when self-pollination may take place. Nipplewort is common by roadsides and in woods and waste places. (*July—October*).

3 Goat's-beard (*Tragopogon pratensis*). This plant is easily recognized by its narrow, grass-like leaves, long, pointed involucral bracts, and fruiting 'clock' formed by beaked achenes with a spreading pappus of branched hairs. Several names, such as Jack-go-to-bed-at-noon and Sleep-at-noon, refer to the flower's habit of closing round about midday. It is common by waysides and in fields and meadows. (*June—July*).

A very similar species, but with purple flowers, Salsify

(*T. porrifolius*), is occasionally found as an escape from gardens, especially near the sea.

4 Coltsfoot (*Tussilago farfara*). A perennial with creeping underground stems. The flowers appear on scaly erect shoots early in the spring and are pollinated by the early bees. The leaves, which develop after the flowers, are large, slightly lobed, and covered with white woolly hairs on the under surface (Fig. 2). Coltsfoot got its Latin name from *tussis*, a cough, because a cough medicine used to be made from the plant. It is common on clay—on waste ground, banks, and ploughed fields, especially where the soil has been disturbed. (*March—April*).

Fig. 2

5 Wall Lettuce (*Lactuca muralis*). A plant with milky juice, growing up to 2 or 3 feet tall. The stem and leaves are smooth; the lower leaves have long stalks, while the upper ones clasp the stem. There are only 4 or 5 florets in each flower-head. The pappus consists of 2 rows of unbranched hairs, the outer being shorter than the inner. Wall Lettuce is fairly common on walls, hedgerows, and in woods and shady places. (*June—September*).

Least Lettuce (*L. saligna*), a much smaller, more slender, and unprickly species, with untoothed, arrow-shaped upper leaves, is found occasionally near the south-east coast. (*July—August*).

Prickly Lettuce (*L. serriola*). A taller plant than Wall Lettuce, with stiff, erect leaves that are prickly on the under side of the midrib (Fig. 3). The florets are pale yellow, often tinged with mauve. The achenes are pale brown when ripe, and with the pappus form a small 'clock'. This is a rather rare plant, though it is on the increase in waste places. (*July—August*).

A stouter species, Greater Prickly Lettuce (*L. virosa*), is darker green with spreading leaves and fruit maroon or nearly black. The numerous small flower-heads with 6 or more florets open in the mornings only. It is also rather rare, but found most often in grassy places and near the sea. (*July—September*).

Fig. 3

LIFE SIZE

1 DANDELION   2 NIPPLEWORT   3 GOAT'S-BEARD
4 COLTSFOOT   5 WALL LETTUCE

COMPOSITAE: DAISY FAMILY (See p. 218)

1 **Marsh Ragwort** (*Senecio aquaticus*). The branched stems grow from 1 to 2½ feet tall. The lowest stem leaves are stalked and either undivided or lobed; while the upper ones are cut into leaflets and partly clasp the stem. The large flower-heads, as with most Ragworts, have spreading outer ray florets and inner disk florets, and all produce smooth fruits (achenes) with white pappus hairs. Marsh Ragwort is common in wet meadows throughout the British Isles. (*June—September*).

Broad-leaved Ragwort or Saracen's Woundwort (*S. fluviatilis*). This very tall but rare plant is found in damp woods and meadows and on riverbanks. It differs from Marsh Ragwort in having long, hairless, undivided, finely toothed leaves, and fewer ray florets—usually only 6 to 8. (*July—September*).

Field Fleawort (*S. integrifolius*). A rare species with hairy, undivided leaves which form a rosette at the base. It is found on chalk downs, mainly in the south. The flower stem grows less than a foot high, and the yellow flower-heads have only one row of bracts and about 13 spreading ray florets. The fruits are hairy. (*May—June*).

2 **Common Ragwort** (*S. jacobaea*). A perennial growing up to 4 feet tall, with a rosette of large, divided leaves at the base which usually dies before the plant flowers. The numerous flower-heads each have about 15 ray florets. The fruits (achenes) differ from other Ragworts in being of two kinds—those of the ray florets are smooth, while those of the disk florets are hairy. This common weed of pastures and waste land often stands conspicuously above the other plants in a meadow because it is not usually eaten by animals. (*June—October*.)

Hoary Ragwort (*S. erucifolius*). Though resembling Common Ragwort, this species has neater, more regularly and deeply cut leaves. Both the leaves, especially the undersides, and the stems, are woolly, and all the fruits are hairy. It is locally common by roadsides and in fields in England and Wales, but rare in Scotland. (*July—September*).

3 **Common Groundsel** (*S. vulgaris*). This very common garden weed, 3 to 12 or more inches high, has 2 rows of bracts below the flower-heads, the outer row of which have black tips. There are no ray florets. Birds eat the hairy fruits, and the plant is gathered as food for cage birds and rabbits. (*January—December*).

Stinking Groundsel or Sticky Groundsel (*S. viscosus*). A locally common species, resembling Common Groundsel but sticky and with a strong, unpleasant smell. The flower-heads usually have some yellow ray florets but these are curled back. The plant grows in dry waste places. (*July—September*).

4 **Oxford Ragwort** (*S. squalidus*). A small plant, about a foot high, with tough, branched stems and smooth leaves, the base of the upper leaves clasping the stem. The 2 rows of bracts below the flower-heads are smooth and black-tipped. The fruits have a long pappus. This plant was probably first introduced into Britain at Oxford, and is now locally common on walls, embankments, and waste places, especially in the south. (*May—October*).

5 **Wood Groundsel** (*S. sylvaticus*). Taller than Common Groundsel and with more hairy stems, this plant has numerous long-stalked flower-heads with very short ray florets that are usually curled back. It is fairly common in Britain in woods and waste places. (*July—September*).

1 Marsh Ragwort   2 Common Ragwort   3 Common Groundsel
4 Oxford Ragwort   5 Wood Groundsel

LIFE SIZE

ORCHIDACEAE: ORCHID FAMILY (See p. 219)

1 Frog Orchid (*Coeloglossum viride*). A plant locally common on hilly pastures, especially on chalky soil, and mainly in the north, and varying in height from 3 inches to about 1 foot. It has 2 lobed tubers (Fig. 1*a*), and the stem, which has 1 or 2 brown scales at its base, is often reddish higher up. The inconspicuous flowers have long green lips and are slightly scented (Fig. 2*a*). (*June—August*).

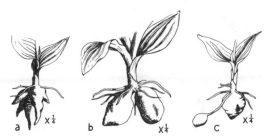

Fig. 1

2 Man Orchid (*Aceras anthropophorum*). This rare orchid, about 6 to 18 inches tall, is easily recognized by the greenish-yellow lip shaped like a man (Fig. 2*b*). The tubers are unlobed (Fig. 1*b*). The Man Orchid grows in grassy places on chalk in south-east England and as far north as Lincoln and Northampton. (*June—July*).

3 Musk Orchid (*Herminium monorchis*). A small plant, growing about 3 to 6 inches high, with rounded tubers (Fig. 1*c*). There are usually only 2 leaves, but there are sometimes 1 or 2 tiny leaves higher up the stem. The flowers are scented, with a 3-lobed lip (Fig. 2*c*). It is rare, being found most often on chalk and limestone in south and east England. (*June—July*).

4 Bird's-nest Orchid (*Neottia nidus-avis*). The name 'bird's-nest' comes from the mass of short swollen roots. The plant is a saprophyte—that is, it has no green leaves but gets its food materials from dead plant or animal matter in the soil. The 2-lobed lip may be straight or curved (Fig. 2*d*). This orchid is found in woods, especially beechwoods, throughout Britain, but is rather rare. (*June—July*).

5 Common Twayblade (*Listera ovata*). Twayblades, as their name indicates, can be recognized by the 2 opposite, oval stem leaves with parallel veins. The greenish-yellow lip of the flower is divided into 2 long narrow segments (Fig. 2*e*). Twayblade is common in woods throughout the British Isles, and grows up to 2 feet tall. (*May—July*).

6 Lesser Twayblade (*L. cordata*). This is like a small version of Common Twayblade (Fig. 2*f*), growing not more than 6 inches high. The leaves are rounded or heart-shaped at the base. It is found in woods and moors and is rare in the south, though less rare in the north. (*June—August*).

7 Coral-root (*Corallorhiza trifida*). The part of the stem below the ground is swollen to form a pale, knobbly rhizome, resembling coral. The plant is a rare saprophyte (*see* No. 4), with no green leaves and inconspicuous flowers, which can be recognized by the crimson markings on the lip (Fig. 2*g*). It grows in damp woods in northern England and Scotland. (*June—July*).

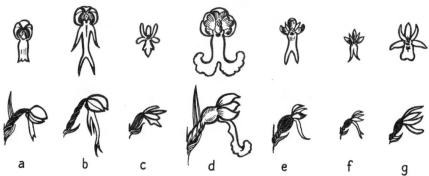

Fig. 2   Orchid Faces.   Front view above, side view below
(*a*) Frog Orchid. (*b*) Man Orchid. (*c*) Musk Orchid. (*d*) Bird's-nest Orchid.
(*e*) Common Twayblade. (*f*) Lesser Twayblade. (*g*) Coral-root

2   3   4   5   6   7

LIFE SIZE

1 FROG ORCHID   2 MAN ORCHID   3 MUSK ORCHID

4 BIRD'S-NEST ORCHID   5 COMMON TWAYBLADE   6 LESSER TWAYBLADE

7 CORAL-ROOT

UMBELLIFERAE: PARSLEY FAMILY (See p. 213)

1 Celery (*Apium graveolens*). A hairless perennial, 1 to 2 feet tall, with a furrowed stem and a strong smell. The flowers in this family (*see also* pp. 86–91) are arranged in groups with all the flower stalks coming from a single point, like the ribs of an open umbrella. Each group is called a simple umbel (Fig. 1*a*), and several umbels are often grouped together to form a compound umbel (Fig. 1*b*). In Celery the umbels are found both at the ends of the stems and in the angle between a leaf and the stem. The stalks of the simple umbels are unequal in length. The plant is fairly common in ditches and on riversides in the south, especially near the sea, but is rarer in the north. It is closely related to the cultivated Celery. (*June—September*).

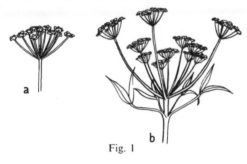

Fig. 1

2 Rock Samphire (*Crithmum maritimum*). A not very common plant of rocks and cliffs by the sea, which grows from 6 to 12 inches high. It has thick and fleshy, pointed leaf segments, which are used in making pickles, and a long sheath encircles the stout stem at the base of each leaf. There are a number of narrow bracts at the base of both the individual umbel stalks and the flower stalks. The tiny yellowish-green flowers develop into comparatively large and ridged purplish-green fruits. The name Samphire comes from the French *herbe de St. Pierre*, or St. Peter's herb. (*June—August*).

3 Alexanders (*Smyrnium olusatrum*). A rather rare biennial 2 to 4 feet high, with solid, furrowed stems, which taste like celery when young. The lower leaves are divided into 3 main segments, each segment being divided into 3 parts once or twice more. The upper leaves are often opposite. There are sometimes a few bracts at the base of individual umbel and flower stalks. The flowers are more yellow than those of most members of this family, and the fruits are nearly black and sharply ribbed (Fig. 2). Alexanders is found on cliffs and waste places near the sea. (*April—June*).

Fig. 2

4 Marsh Pennywort or White-rot (*Hydrocotyle vulgaris*). This small, creeping or floating plant, which is sometimes put in a family by itself, differs from other Umbellifers in having undivided round leaves, with the leaf stalk attached at the centre of the leaf blade. The leaves are more like those of Wall Pennywort (*see* p. 53) but the flowers are quite different. The tiny, greenish-white flowers are in clusters on stalks shorter than the leaf stalks. Marsh Pennywort is common in bogs and marshes throughout the British Isles. (*June—August*).

5 Fennel (*Foeniculum vulgare*). A strong-smelling perennial, 2 to 4 feet high, with a shiny stem and very narrowly divided leaves with sheath-like bases. There are usually no bracts below the flowers and umbels. The yellowish petals have inwardly curled points. The oval, flat, and ridged fruits are reddish-brown. Fennel is a not very common plant, growing in waste places, especially on sea cliffs. It is not usually found far north. It is grown in gardens as a herb, the dried leaves being used to flavour sauces to serve with fish. (*July—October*).

HALF LIFE SIZE

1 CELERY   2 ROCK SAMPHIRE   3 ALEXANDERS
4 MARSH PENNYWORT   5 FENNEL

47

1 Lady's Mantle (*Alchemilla vulgaris*, family *Rosaceae*). A perennial plant with a stout rootstock which varies a good deal, chiefly in hairiness and in leaf-shape. The lowest leaves have long stalks, while the stem leaves have very short stalks, and at the point where these join the stem, small leaf-like structures called stipules. The tiny flowers have no petals, but 4 small bracts are arranged alternately with 4 sepals, the 4 stamens being opposite the bracts (Fig. 1). The single style has a rounded stigma. Lady's Mantle is found on grassland throughout the British Isles, though not very often in the south and east. (*June—September*).

Fig. 1

Alpine Lady's Mantle (*A. alpina*). This is like a small Lady's Mantle, but with leaves divided almost or completely to the base into 5 to 7 lobes, dark green on the upper surface and silvery underneath (Fig. 2a). The plant is locally common on mountains in the north. (*June—August*).

Parsley Piert (*Aphanes arvensis*). A small hairy annual easily distinguished from Lady's Mantle because each leaf is very much divided (Fig. 2b). The flowers are in groups opposite the leaves, and are hidden by leafy stipules. There are only 1 or 2 stamens. Parsley Piert is common in fields and dry waste places. (*May—September*).

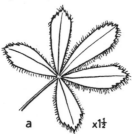

a   x1½   b   x1½

Fig. 2

2 Wild Madder (*Rubia peregrina*, family *Rubiaceae*). An evergreen with straggling stems up to 3 or 4 feet long. The prickles on the 4 angles of the stem, the leaf edges, and the underside of the midrib are a distinctive feature. There is no calyx, a corolla with usually 5 lobes, and 5 stamens. The fruit is a black berry. It is a rather rare plant of hedgebanks and scrub in the south and west. (*June—August*).

The dye madder is obtained from the roots of another species (*R. tinctorum*), with bright yellow flowers and a reddish-brown berry, which is sometimes found growing in England.

3 Knawel (*Scleranthus annuus*, family *Illecebraceae*). An inconspicuous, low, spreading, annual plant, with very small petal-less flowers. The stamens are shorter than the sepals, and there are 2 styles. It is fairly common in dry sandy fields and waste places. (*July—September*).

The Perennial Knawel (*S. perennis*) is found occasionally in dry sandy fields in Norfolk and Suffolk, and is recognized by its woody base and white-edged sepals.

4 Wild Mignonette (*Reseda lutea*, family *Resedaceae*). This plant is not as sweet scented as the garden variety. It grows from 1 to 2 feet tall, the flower stem being solid and rough. The flowers usually have 6 sepals and 6 petals. It is only locally common in waste places and ploughed fields on chalky soil. (*June—August*).

Dyer's Rocket or **Weld** (*R. luteola*). This biennial is taller than Wild Mignonette, and has a hollow, unbranched or slightly branched stem. The narrow leaves with wavy margins are not divided into leaflets (Fig. 3). There are usually 4 sepals and 4 petals. It is more common than the last species, especially in the south.

Fig. 3

5 Townhall Clock or Moschatel (*Adoxa moschatellina*, family *Adoxaceae*). The name Townhall Clock refers to the arrangement of the clusters of flowers, usually with one pointing upwards and four facing outwards in different directions. The other name, coming from a word meaning 'musk', describes the smell. There are 2 or 3 sepals, 4 or 5 petals, and twice as many stamens as petals. The plant is common in woods, hedgerows, and damp, shady places. (*March—May*).

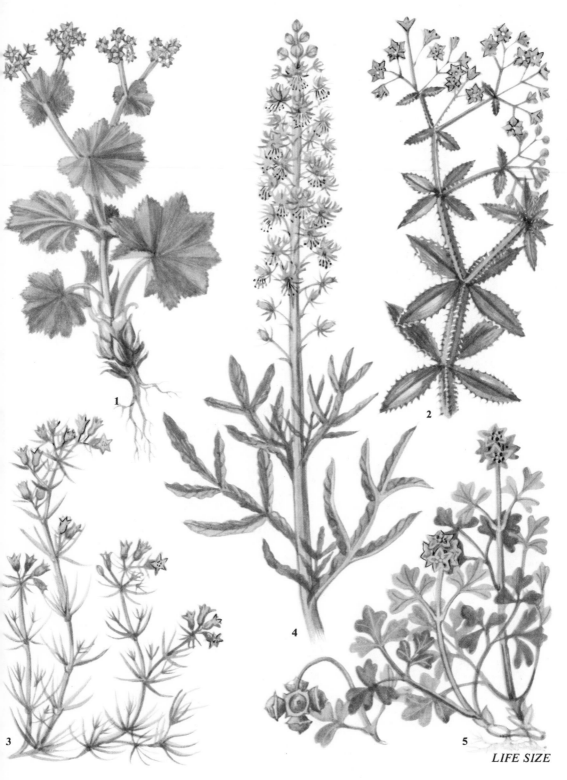

1 Lady's Mantle   2 Wild Madder   3 Knawel
4 Wild Mignonette   5 Townhall Clock

LIFE SIZE

1 Traveller's Joy (*Clematis vitalba*, family *Ranunculaceae*). A climbing plant, often completely covering hedges and bushes. It climbs by twisting its leaf stalks round the stems of other plants. The small, greenish-cream flowers attract flies and bees by their sweet, rather sickly scent. The masses of feathery fruits, conspicuous in autumn, give the plant the name Old Man's Beard (Fig. 1). This is one of the characteristic shrubs of chalky soil, but is far more common in the south than in the north. (*July—August*).

Fig. 1

2 Mistletoe (*Viscum album*, family *Loranthaceae*). The Mistletoe is the only British member of its family, and is found most commonly in the south. It is partly a parasite, that is, it gets most of its food materials from the trees on which it grows, so weakening them. The stiff, branched stems hang in bunches from high up on a tree, often an apple or poplar tree. The sticky white berries ripen in the winter and attract birds, especially thrushes—hence missel-thrush. The seeds stick to the birds' beaks and are often wiped off on to a branch of another tree, thus spreading the Mistletoe from tree to tree. The Mistletoe was connected with very old pagan customs and beliefs, and is still today supposed to be lucky—the reason for kissing under the Mistletoe bough. (*February—May*).

3 Green Hellebore (*Helleborus viridis*, family *Ranunculaceae*). A plant from 4 to 8 inches high at the flowering period, but the leaves continue to grow after the flowers are over. A plant usually bears 3 or 4 flowers with green petal-like sepals, and only rudimentary petals (Fig. 2). Early bees can be seen visiting the 9 to 12 tiny tubular nectaries in the flowers. This is a rare perennial plant of damp meadows, woods, and copses on chalky soil, often found among Dog's Mercury (*see* p. 63). It is closely related to garden Christmas Roses. (*March—April*).

Fig. 2

Stinking Hellebore or Bear's-foot (*H. foetidus*). A stouter, taller species than Green Hellebore with a strong, unpleasant scent. Each plant bears more flowers, and the petal-like sepals curve inwards, and often have a purplish border. It is a rare woodland plant of chalky soils. (*February—April*).

4 White Bryony (*Bryonia dioica*, family *Cucurbitaceae*). This climbing plant is the only British member of a family which includes garden Marrows and Cucumbers. The tendrils, which are modified branch stems opposite the leaves, grow in springlike coils which, on touching another object, curl round it, forming a strong, flexible attachment. In this way a plant with a weak main stem can reach the tops of hedges. The male and female flowers are produced on separate plants. The berries turn gradually from green through yellow and orange to red as they ripen. White Bryony, though fairly common in southern England, becomes much less common further north. (*May—July*).

5 Black Bryony (*Tamus communis*, family *Dioscoreaceae*). A climbing plant, the only British member of the Yam family and, in spite of its name, not related to White Bryony. It climbs, not by tendrils but by twining thin, smooth stems clockwise around the stems of hedgerow plants. The inconspicuous flowers have 6 sepal-like segments, and the male and female flowers are on different plants. The plant is easily recognized by its shiny broad leaves and scarlet berries. The name 'black' refers to its black tubers. Many tropical members of the family are cultivated for these starchy tubers. (*May—July*).

THREE-QUARTERS LIFE SIZE

1 TRAVELLER'S JOY   2 MISTLETOE   3 GREEN HELLEBORE
4 WHITE BRYONY   5 BLACK BRYONY

1 **Herb Paris** (*Paris quadrifolia*, family *Liliaceae*). A perennial growing about 1 foot high, with a creeping rhizome. It is easily recognized by the whorl of usually 4 leaves and the strange flowers, which mostly have 4 narrow sepals, 4 broader petals, 8 yellowish stamens, and 4 purple carpels. This plant is fairly common in damp woods on chalk, but is not found in Ireland. (*April—June*).

2 **Wall Pennywort** (*Umbilicus rupestris*, family *Crassulaceae*). A perennial varying in height from a few inches to well over 1 foot in damp shady places. The leaf-stalk is attached to the centre of the leaf-blade. The bell-shaped flowers have 5 small sepals and 5 teeth at the end of the corolla tube. It is fairly common on walls and rocks, especially in the west. (*June—August*).

Pondweeds (*Potamogeton*, family *Potamogetonaceae*). These are perennial water plants of which there are many species, difficult to identify because they vary considerably according to whether they are growing in deep or shallow, still or fast-moving water, and because they hybridize easily with each other. In general they grow in clumps spreading 3 to 6 or more feet, with submerged leaves and some also with floating leaves. The flowers are small and petal-less, with 4 sepals, stamens, and stigmas, and the fruit is made up of 4 nutlets. (*May—August*).

3 **Broad-leaved Pondweed** (*P. natans*). This species, which is shown on p. 53, has opaque floating leaves on thick stalks and long, narrow stipules. The underwater leaves are often reduced merely to stalks. It is common in still water, often covering ponds or ditches.

Another less common species (*P. gramineus*) has few floating leaves and narrow, unstalked submerged leaves. It has special buds, which survive the winter, at the ends of creeping stems.

Shining Pondweed (*P. lucens*) is distinguished by its oblong, shiny, submerged leaves, often 6 inches long and with 10 or more parallel veins. The stipules at the base of the leaf are long and shaped like the keel of a boat. It is fairly common in still or slow-moving water.

Another common species of ponds and streams is Perfoliate Pondweed (*P. perfoliatus*). Its leaves, smaller than those of the last species and entirely clasping the stem, are all submerged. Curled Pondweed (*P. crispus*), also common, has submerged leaves with wavy margins.

Some Pondweeds, such as Small Pondweed (*P. berchtoldii*) and Fennel-leaved Pondweed (*P. pectinatus*), are much slenderer and have almost grass-like submerged leaves.

4 **Mare's-tail** (*Hippuris vulgaris*, family *Haloragaceae*). The leafy shoots arise from a creeping rhizome, and are usually partly submerged in fairly fast-moving water. The tiny petal-less flowers have only one stamen with red anthers. Mare's-tail is a widespread but only locally common plant of lakes, streams, and ditches. (*June—July*).

5 **Water Starwort** (*Callitriche stagnalis*, family *Callitrichaceae*). A common, slender plant, growing in still or slow-moving water or lying on wet mud. There are separate male and female flowers; the male flower has a single yellow stamen, the female a 4-lobed ovary and 2 styles, which bend downwards when the plant is fruiting. The lobes of the fruit are winged. (*April—September*).

There are several other species, one of which, *C. intermedia*, can be distinguished by its long, narrow lower leaves, notched at the ends, and unwinged fruits.

6 **Horn-wort** (*Ceratophyllum demersum*, family *Ceratophyllaceae*). A completely submerged waterweed with stiff, forked leaves. The ripe fruit has 2 spines at the base. This perennial is locally common in ponds and ditches in England and Ireland, but rare in Wales and Scotland. (*June—September*).

A rarer species, *C. submersum*, is without spines to the fruit. It is most often found near the sea in the south and east.

7 **Canadian Pondweed** or Water-thyme (*Elodea canadensis*, family *Hydrocharitaceae*). A much-branched underwater plant that survives the winter as special swollen buds. The male and female flowers grow on separate plants, the male being very rare. The female flowers, which grow on stalk-like threads, have 3 petals, and 3 feathery stigmas. The plant was introduced in Ireland from North America and is now common in fresh water throughout the British Isles. (*July—September*).

8 **Spiked Water-milfoil** (*Myriophyllum spicatum*, family *Haloragaceae*). The lower flowers are female, with 4 small petals, and the upper ones male, with larger red petals, soon dropping. The bracts below the whorls of flowers are usually shorter than the flowers. The plant is locally common in ditches and ponds. (*June—August*).

There are two other rather less common Water-milfoils. *M. verticillatum* has the bracts below the whorls of flowers longer than the flowers; and *M. alterniflorum* is a more slender plant, with the upper flowers arranged singly or in opposite pairs.

LIFE SIZE

1 HERB PARIS   2 WALL PENNYWORT   3 BROAD-LEAVED PONDWEED

4 MARE'S-TAIL   5 WATER STARWORT   6 HORN-WORT

7 CANADIAN PONDWEED   8 SPIKED WATER-MILFOIL

CHENOPODIACEAE: GOOSEFOOT FAMILY (See p. 211)

1 Common Orache (*Atriplex patula*). An annual with much-branched, stiff, spreading stems, growing up to 3 feet or more in height. The leaves are sometimes covered with a mealy substance, and the lower leaves are usually opposite. Separate male and female flowers grow on the same plant, the male having 5 sepals and 5 stamens, while the female are without sepals and petals but are surrounded by 2 triangular bracts, which get bigger as the fruit ripens (Fig. 1). Orache is common in waste places and cultivated fields throughout the British Isles. (*July—September*).

A very similar species to Common Orache, Broad-leaved Orache (*A. hastata*), is more erect and has leaves that are smooth, or mealy only underneath. The leaf-blade does not taper gradually to the leaf-stalk, as it does in Common Orache. The plant is locally common in waste places by the sea.

Fig. 1

Frosted Orache (*A. sabulosa*). This small annual of sea-shores has branched, yellowish or reddish stems up to 12 inches high, and whitish leaves which are mealy on both sides. (*August—September*).

2 Grass-leaved Orache or Shore Orache (*A. littoralis*). This rather rare species of salt-marshes differs from other Oraches in having all linear leaves, usually covered with meal. The much-branched stems are from 2 to 3 feet tall. The spikes of flowers have leaves at the base only. The 2 triangular bracts round the fruits are rough and bumpy. (*July—September*).

Goosefoots (*Chenopodium*). About 18 species of Goose-foot may be found in Britain, many of them probably brought in with cargoes from abroad and now naturalized on rubbish dumps and waste places. They are similar to the Oraches in general appearance but usually have both male and female organs (stamens and pistil) in the same flower and 2 to 5 sepals (Fig. 2).

It is often impossible to name a plant for certain without examining the seedcoat under the microscope, so only the most distinct species are described here.

Fig. 2

3 Stinking Goosefoot (*C. vulvaria*). A low annual, up to 12 inches high, with a strong smell of bad fish. The underside of the leaves is mealy. It is a rather rare plant of salt-marshes, shingle, and waste places, absent from Ireland and the north of Scotland. (*July—September*).

4 Red Goosefoot (*C. rubrum*). An often somewhat reddish annual, with nearly smooth stems, up to 2 feet high. There may be leaves among the crowded clusters of flowers. The seeds are reddish-brown and rough. It is fairly common in England, on waste or cultivated ground and near the sea, but rare elsewhere in the British Isles. (*August—October*).

5 Good King Henry (*C. bonus-henricus*). A perennial with a thick, fleshy rootstock and large, dark-green, un-toothed leaves. The flower clusters are mostly at the end of the stem and without leaves. The stigmas are long, and the reddish-brown seeds are not enclosed by the sepals. It is a fairly common plant of roadsides and fields in England and Wales, becoming rarer in the north. (*May—August*).

6 White Goosefoot or Fat Hen (*C. album*). A rather mealy annual which was at one time eaten as a vegetable, like Spinach. The lower leaves are toothed, but the upper often untoothed. The flowers are in spikes at the end of the stem and also in the angles between the stem and leaves. The smooth seeds are enclosed by sepals. It is very common on waste and cultivated ground throughout the British Isles. (*July—September*).

LIFE SIZE

1 COMMON ORACHE   2 GRASS-LEAVED ORACHE   3 STINKING GOOSEFOOT

4 RED GOOSEFOOT   5 GOOD KING HENRY   6 WHITE GOOSEFOOT

CHENOPODIACEAE: GOOSEFOOT FAMILY (See p. 211)

1 Glasswort (*Salicornia stricta*). The fleshy leaves of this annual are fused together in pairs to give a jointed look to the stem. The flowers are in threes at joints at the ends of the stems. Glasswort, so called because its ashes were at one time used in glass-making, is fairly common in salt-marshes. (*August—September*).

There are several other species of Glasswort, all with the same general appearance and all found in salt-marshes. *S. ramosissima* has more numerous branches and shorter flowering spikes, with the middle flower about twice as big as the other two. A perennial species (*S. perennis*) has woody, creeping stems and a 2-lobed stigma. *S. prostrata* lies on the mud and is often reddish. The short upright stems of the rare *S. disarticulata* carry only one flower at a joint and become brownish-yellow with pink tips to the segments, which break apart before the seeds are ripe.

2 Sea Purslane (*Halimione portulacoides*). A woody perennial from 1 to 2 feet in height, which is fairly common in salt-marshes, except in the north. The leaves are covered with grey 'meal', which is formed by bladder-like hairs. There are separate male and female flowers on the same plant. The fruits are unstalked. (*July—September.*)

A very rare species (*H. pedunculata*) is found in east-coast salt-marshes. It is a small annual with all its leaves alternate and with stalked fruits.

3 Saltwort (*Salsola kali*). A low, prickly, creeping plant, usually much branched. The flowers have 5 sepals, no petals, 5 stamens, and 2 or 3 styles (Fig. 1). Saltwort is fairly common round sandy coasts and salt-marshes. At one time its ash was used in making soap. (*July—September*).

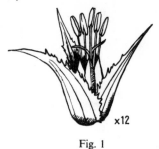

Fig. 1

A more bushy and erect species, Russian Thistle (*S. pestifera*), is still very rare, though on the increase.

It has longer, darker green, and less spiny leaves, and grows from 6 to 18 inches tall.

4 Wild Beet (*Beta vulgaris*). A variable plant, with upright or spreading, smooth or slightly hairy stems, about 2 feet tall. The large lower leaves often form a rosette at the base of the plant. The 5 green sepals grow bigger as the fruit ripens (Fig. 2). Wild Beet is common on sea-shores throughout the British Isles, and exists in two forms, one of which is reddish in colour and the other has a very swollen root. The cultivated Sugar-beet, Beetroot, and Spinach Beet are varieties of the second form. (*July—September*).

Fig. 2

5 Seablite (*Suaeda maritima*). A low annual with many branches and fleshy leaves. The flowers are arranged singly in the angle between the leaf and the stem. There are 5 stamens opposite the 5 sepals, and 2 styles (Fig. 3a). This is a fairly common plant of salt-marshes and sea-shores. (*July—September*).

Shrubby Seablite (*S. fruticosa*) is a rare evergreen shrub up to 3 feet tall, of sea coasts above high-water mark in south and east England. The flowers have 3 styles, joined together at the bottom, and divided into lobes at the top (Fig. 3b).

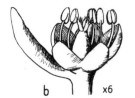

Fig. 3

LIFE SIZE

1 GLASSWORT   2 SEA PURSLANE   3 SALTWORT
4 WILD BEET   5 SEABLITE

POLYGONACEAE: DOCK FAMILY (See p. 213)

1 Common Sorrel (*Rumex acetosa*). An erect common perennial, up to 2 or 3 feet high, with nearly smooth stem and leaves. The leaves, which have a sharp taste, are sometimes added to salads. They are arrow-shaped with downward-pointing lobes at the base. Male and female flowers grown on different plants (**a** and **b** in the illustration), and the flower spikes gradually turn red. The female flowers, which are wind pollinated, have 3 outer and 3 inner sepal-like segments (Fig. 1*a*). After flowering time, the former turn down round the flower-stalk, and the latter surround the fruit (Fig. 1*b*). Common Sorrel grows in meadows throughout the British Isles. (*May—July*).

a   b

Fig. 1

2 Mountain Sorrel (*Oxyria digyna*). This small, hairless perennial, about 6 inches or less in height, has round or kidney-shaped leaves, and leafless spikes of red-edged, greenish flowers. There are 4 sepal-like segments, 2 of which surround the winged fruit, and 6 stamens, and 2 stigmas. It is locally common in damp places on mountains in the north. (*June—August*).

3 Sheep's Sorrel (*Rumex acetosella*). This plant is smaller and more slender than Common Sorrel. Its leaves, which are all stalked, have spreading lobes at the base. The male and female flowers are on separate plants, and, like those of Common Sorrel, turn red as they mature. It is common on heaths and grassy places, except on chalky soil. (*May—August*).

4 Sharp Dock or Clustered Dock (*R. conglomeratus*). A slender perennial 1 to 2 feet tall, with a wavy stem and spreading branches. The leaves are long and pointed and rounded at the base, and the whorls of flowers are spaced out along the stems, with leafy bracts between them. Each of the 3 perianth segments surrounding the fruits has an elongated swelling or wart. This Dock is common by waysides and in damp places throughout most of Britain, but rare in Scotland. (*June—August*).

The Wood or Red-veined Dock (*R. sanguineus*) is a common southern species, differing from the Sharp Dock in having a nearly straight stem, more upright branches, and fewer leaves between the whorls of flowers. Only one of the perianth segments has a wart. The rarer Golden Dock (*R. maritimus*) has narrow yellowish-green leaves and dense whorls of flowers. The whole plant turns bright yellow when fruiting. It is found locally on damp or muddy ground in England and Ireland.

5 Broad Dock (*R. obtusifolius*). A large plant, up to 3 feet tall, which is very common by waysides and on waste places. There are often hairs underneath the large oblong leaves. The perianth segments have prominent teeth and only one has a swelling or wart. The fresh, broad leaves were often used by country people to wrap round butter to keep it cool, and they are supposed to lessen the pain of Nettle stings. (*June—September*).

6 Water Pepper (*Polygonum hydropiper*). A slender plant, up to 2 feet high, with sheaths round the stem above the leaf-bases. The leaves have a peppery taste and a juice which makes the skin itch. The 3 sepals are covered with yellow dots (glands). The fruits are a dull blackish colour. The plant is common in ditches and other wet places throughout the British Isles, except north Scotland. (*July—September*).

A smaller species, Small Persicaria or Least Water Pepper (*P. minus*), has pinkish flowers without the yellow dots on the sepals, and shiny black fruits. It is locally common in wet places.

7 Curled Dock (*Rumex crispus*). A tall Dock, up to 3 feet tall, which is distinguished by the wavy, curled leaf-edges. The perianth segments have no teeth and each usually has a wart. This Dock is very common on waste places, waysides, and cultivated land. (*June—September*).

Another species, Butter Dock (*R. longifolius*), differs from Curled Dock in having broader leaves, denser and less branched flower spikes, and no warts.

Great Water Dock (*R. hydrolapathum*). This is the largest British Dock and is fairly common in wet places, especially by rivers. It grows to about 5 feet high, and the lower leaves are up to 2 feet long. (*July—September*).

8 Black Bindweed (*Polygonum convolvulus*). A twining plant with a rather mealy, angular stem, which may grow as much as 3 feet long. It twines clockwise, whereas true Bindweeds twine anti-clockwise. Its leaves resemble those of true Bindweeds, but its flowers are entirely different. It is a common weed of fields and waste places. (*June—October*).

1a

1b

4a

5a

2

3

1

4

5

6

7

8a

8

ONE-EIGHTH LIFE SIZE

1 COMMON SORREL   2 MOUNTAIN SORREL   3 SHEEP'S SORREL
4 SHARP DOCK   5 BROAD DOCK   6 WATER PEPPER
7 CURLED DOCK   8 BLACK BINDWEED

59

1 Hoary Plantain (*Plantago media*, family *Plantaginaceae*). A perennial with hairy, ribbed leaves which taper to a very short stalk and, as with all Plantains, form a rosette on the ground. The spike of minute flowers of all Plantains owes its colour to the prominent stamens and is borne on an unbranched leafless stem. Hoary Plaintain, which may grow up to 12 inches high, is common on chalky pastures in the south, though rarer in the north. (*May—August*).

2 Greater Plantain or Ratstail (*P. major*). The strongly veined leaves, which may be smooth or hairy, differ from those of Hoary Plantain in having longer stalks. The plant is coarser and may grow to 18 inches in height. The flower spike is as long as or longer than its stalk. This Plantain is very common by roadsides, in meadows, and on cultivated ground, being a tiresome weed in lawns. (*May—September*).

3 Buck's-horn Plantain (*P. coronopus*). This is the only Plantain in which the leaves are divided into leaflets with one vein. The plant is often very small. It is most common in sandy places near the sea. (*May—July*).

Sea Plantain (*P. maritima*). A small, rather woody-stemmed plant, with narrow, usually smooth, fleshy, and veined leaves, and longish, brownish flower spikes with pale yellow stamens. The plant grows in salt-marshes, by the sea, and on mountains, but is less common than Buck's-horn Plantain. (*June—August*).

4 Ribwort Plantain (*P. lanceolata*). The erect leaves are ribbed with 3 to 6 veins. The flower stalk is deeply furrowed, and the flower spikes are shorter and much more compact than those of other Plantains. This is a very common plant of roadsides and meadows and grows from 6 inches to 2 feet tall. (*April—September*).

Shoreweed (*Littorella uniflora*). A small plant, about 3 inches high, with a tuft of bright green, grass-like leaves (Fig. 1). It has creeping stems rooting in mud, sand, or shallow water at the edges of ponds and lakes. There are separate male and female flowers. One male flower with 4 prominent stamens is borne at the end of each flower stalk, while the female flowers are hidden in the leaves at the base of the flower stalks, with only the long slender styles visible. It is locally common throughout the British Isles. (*June—August*).

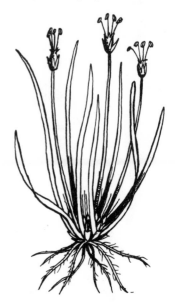

Fig. 1

5 Mugwort (*Artemisia vulgaris*, family *Compositae*). A perennial, growing 2 to 3 feet tall, with an aromatic scent. The leaves are dark green on top and covered with white hairs underneath. The spikes of numerous flower-heads are held nearly erect, and the bracts round the flower-heads are woolly. Mugwort is common in waste places and hedgerows throughout the British Isles. (*July—September*).

There is a smaller, less woolly species with finely divided leaves (*A. campestris*), found sometimes in grasslands in Breckland, East Anglia. The flower-heads are rounder and more yellow, and the plant is scentless.

6 Sea Wormwood (*A. maritima*). The woody, much-branched stems and the much-divided leaves are covered with white woolly hairs. The numerous small flower-heads have yellowish or reddish florets. The plant grows about 1 foot high, being slenderer and smaller than Wormwood (*see* p. 32). It is locally common in salt-marshes round the British Isles, except in the extreme north. It has a pleasant aromatic scent. (*July—September*).

1   1a   2a   3   3a   2   4a   4   5a   5   6   6a

LIFE SIZE
AND (**a**) *ONE-QUARTER LIFE SIZE*

1 HOARY PLANTAIN	2 GREATER PLANTAIN	3 BUCK'S-HORN PLANTAIN
4 RIBWORT PLANTAIN	5 MUGWORT	6 SEA WORMWOOD

1 Sun Spurge (*Euphorbia helioscopia*, family *Euphorbiaceae*). A smooth annual, up to a foot or more high, which, like all Spurges, contains a milky juice. The stem is usually unbranched below the flower-cluster, which has 5 main branches. The leaves and the bracts at the base of the flower-stalks are alike, except that the bracts are often tinged with yellow. Each 'flower' has several minute stamens, each representing a separate male flower and surrounding one female flower with a 3-lobed ovary and 3 styles (Fig. 1). Sun Spurge is a common weed of gardens, fields, and waste places throughout the British Isles. (*April—November*).

Fig. 1

Petty Spurge (*E. peplus*). This common annual is smaller and more slender than Sun Spurge and has oval, untoothed leaves and bracts. The 'flowers' are surrounded by bracts joined together to form an involucre, with crescent-shaped glands round the edges. These glands produce nectar, which attracts insects to the 'flowers'. This plant grows throughout Britain in much the same places as Sun Spurge. (*May—November*).

Dwarf Spurge (*E. exigua*). This small annual plant differs from the other Spurges described here in having very narrow, unstalked leaves (Fig. 2). Though common in most parts of Britain, it is rare in northern Scotland and western Ireland. (*June—October*).

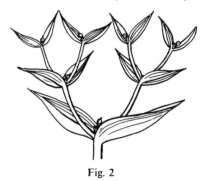

Fig. 2

2 Wood Spurge (*E. amygdaloides*). A downy perennial, 1 to 2 feet tall. Flowers do not appear until the second year,

when the first year's leaves are crowded together in the middle of the stem. The bracts are joined together in pairs below the groups of 'flowers' and each 'flower' cup has crescent-shaped glands round the edge. This is far more common in woods and copses in the south than further north. (*March—June*).

Sea Spurge (*E. paralias*). The pale green leaves of this perennial are thick, fleshy, and crowded together on the stems. The flower-cluster has 3 to 6 main branches, and the rounded bracts are also thick and fleshy. It is locally common on sand-dunes and sea-shores. (*June—October*).

Another species of much the same habitat (*E. portlandica*) is less tall and robust than Sea Spurge and rather greyer in colour. The leaves are small, lanceolate, pointed, and rather leathery.

3 Lords-and-ladies or Cuckoo-pint (*Arum maculatum*, family *Araceae*). This common plant of woods and hedgerows is easily recognized by its leaves, which appear very early in the spring, its flower-spike (spadix), which is purple or greenish-white, and its brilliant poisonous fruits. The flowers are below the purple club-shaped part of the spadix, the female flowers being below the male flowers. There is a tuft of stiff downward-pointing hairs above the male flowers, which often trap small flies visiting the plants. The flies then crawl over the female flowers, bringing pollen from another spadix. (*April—June*).

4 Dog's Mercury (*Mercurialis perennis*, family *Euphorbiaceae*). A rather hairy perennial with upright, unbranched stems produced from a creeping rhizome. Male and female flowers are on separate plants, and pollen is carried by the wind. The female flower (4*a*) has a calyx with 2 or 3 teeth, no petals, and an ovary covered with bristles. The fruit (4*b*) is a 2-lobed capsule with a seed in each lobe. Dog's Mercury is very common in woods and hedges in most parts of the British Isles, and is one of the few plants that flourish in beechwoods. (*February—May*).

Annual Mercury (*M. annua*). This species is a paler green than Dog's Mercury and has smooth, branched steams. The female flowers have very short stalks or none. It grows in waste places, especially in the south, and is locally fairly common. (*June—October*).

1 SUN SPURGE   2 WOOD SPURGE   3 LORDS-AND-LADIES

4 DOG'S MERCURY

LIFE SIZE

1 **Pellitory-of-the-wall** (*Parietaria diffusa*, family *Urtica-ceae*). A perennial, with rounded stems, sometimes up to 3 feet in height. The branches often spread outwards or grow along the ground. The clusters of flowers, surrounded by a few bracts, may be (*a*) female (ovary only), (*b*) male (stamens only), or (*c*) with both stamens and ovary in one flower (Fig. 1). The stigmas are either red or white. Pellitory is locally common on walls and stony places, but rarer in Scotland. (*May—October*).

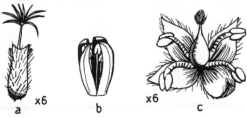

Fig. 1

2 **Stinging Nettle** (*Urtica dioica*). A very common and widespread perennial weed with yellow creeping underground stems and tough roots. The angular, up-right stems grow to 3 feet high or more. Stems and leaves are covered with stinging hairs, the tips of which break off when touched, releasing an acid which causes the stinging sensation. Some plants have (*a*) female flowers, others (*b*) male flowers only (Fig. 2); pollen is carried from plant to plant by the wind. In country districts, the young Nettle shoots are sometimes cooked and eaten, like Spinach, and during the Second World War large quantities of Nettles were gathered and the chlorophyll in them extracted for use in medicines and to make a green dye used in camouflage. (*June—September*).

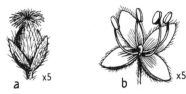

Fig. 2

Small Nettle (*U. urens*). This is an annual, similar in appearance to the Stinging Nettle, but smaller, with the lower leaves shorter than their stalks, and with the male and female flowers on the same plant. It is locally common on waste and cultivated land. (*May—September*).

3 **Ivy** (*Hedera helix*, family *Araliaceae*). A woody ever-green, common on walls and in woods and hedges throughout the British Isles. It climbs by small roots developed along the stem, which attach themselves firmly to a tree trunk or wall by a cement-like substance. There are two sorts of leaves: those on the climbing or creeping stems have 3 to 5 triangular lobes, while those on the flowering stems are unlobed. The flowers are pollinated by flies and wasps seeking nectar. The fruit is a black berry. (*September—November*).

4 **Bastard Toadflax** (*Thesium humifusum*, family *Santala-ceae*). A small, spreading perennial, partly a parasite, that is, it gets some of its food material by sending out suckers from its roots to those of other plants. There are 5 stamens opposite the 5 greenish-white perianth segments. The fruit is a small green nut with the peri-anth segments still attached on the top. This is a rare plant of grassland on chalky soil in southern England. (*May—August*).

5 **Hop** (*Humulus lupulus*, family *Cannabinaceae*). This climbing perennial twists its rough stems in a clockwise direction round other plants. The male and female flowers are developed on different plants. The male flowers (see bottom of fig.) consist of 5 perianth seg-ments and 5 stamens, while the female flowers are in the pale green cones of bracts, with 2 flowers at the base of each bract. It is common in hedges and open woods through most of Britain, and is grown in certain counties, such as Kent and Worcestershire, for use in brewing. (*July—September*).

1 PELLITORY-OF-THE-WALL   2 STINGING NETTLE   3 IVY
4 BASTARD TOADFLAX   5 HOP

LIFE SIZE

1 Wood Anemone (*Anemone nemorosa*, family *Ranunculaceae*). A perennial plant from 3 to 6 inches high. There are from 5 to 7, usually 6, petal-like segments, which often have a pink or purplish tinge. The fruits are achenes. The plant dies down each year, but the underground stem, called a rhizome, survives the winter. Large masses of these attractive flowers can be found on hedgebanks and carpeting woods in the spring. (*March—May*).

2 Water Crowfoot (*Ranunculus aquatilis*). There are altogether some thirteen British species of Water Crowfoot, nearly all with white Buttercup-like flowers and many with two different types of leaf — broad and often lobed ones that float on the surface of the water, and fine, thread-like ones that grow submerged. They are all to be found in streams and ponds, some preferring mud, others still water, and others running water. The different species can be distinguished by the shapes of the leaves and size of the flowers.

One variety of the plant illustrated here has its floating leaves cut deeply into broad straight-sided segments, and the submerged leaves collapse when they are taken out of the water. Another variety has its floating leaves less deeply cut into rounded lobes, and the submerged leaves remain stiff when taken from the water.

Another species, Ivy-leaved Water Crowfoot (*R. hederaceus*), has Ivy-shaped floating leaves with the lobes broadest at the base, and usually no submerged leaves, for the plant grows more often in wet mud than in water. The flowers are small with petals and sepals of about the same length.

A less common species, *R. lenormandi*, has larger flowers with petals twice as long as the sepals, and the lobes of the leaves narrowest at the base.

Thread-leaved Crowfoot (*R. trichophyllus*), which is fairly common in ponds, has small flowers and no floating leaves. Another species which lacks floating leaves is River Crowfoot (*R. fluitans*). The submerged leaves are like long tassels, and the flowers are large, about 1 inch across, with 5 to 10 overlapping petals. It is found in running water.

The various species of Water Crowfoot flower from *April to September*.

3 Baneberry or Herb Christopher (*Actaea spicata*). A number of small white flowers are arranged along a smooth flowering stem, which grows from 1 to 2 feet tall. The 4 sepals and 4 petals are alike. The fruits, which are poisonous, ripen from green to shiny black berries. The rhizomes were at one time used in cures for skin diseases and asthma. This is a rare plant of limestone, found locally in woods in the north. (*May— June*).

4 White Waterlily (*Nymphaea alba*, family *Nymphaeaceae*). This is less common than the Yellow Waterlily, though it can be found on lakes and ponds in all parts of the British Isles. Both the large flowers and the nearly circular leaves float on the surface of the water. The fruit is a capsule that ripens under water and splits to release the seeds (Fig. 1). These have air in their seedcoats, enabling them to float. (*June— August*).

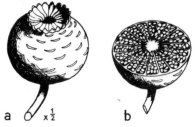

a $\times \frac{1}{2}$ b

Fig. 1

(*a*) Whole fruit   (*b*) Fruit cut in half

LIFE SIZE

1 WOOD ANEMONE   2 WATER CROWFOOT   3 BANEBERRY
4 WHITE WATERLILY

CRUCIFERAE: CABBAGE FAMILY (See p. 209)

1 Watercress (*Nasturtium officinale*). Watercress is cultivated in beds in shallow rivers; it also grows wild in streams and ditches throughout Britain. The stems creep on mud or float in shallow running water. The hairless leaves, which remain green throughout the year, end in an odd leaflet, usually bigger than the others. There are 4 long and 2 short stamens, as in most members of the Cabbage family. (*May—October*).

Another common Watercress (*N. microphyllum*) has leaves that turn brown in autumn. The end leaflet is hardly bigger than the rest.

2 Jack-by-the-hedge or Garlic Mustard (*Alliaria petiolata*). This plant smells of garlic when rubbed. The usually unbranched stem grows straight upwards from a rosette of long-stalked leaves. The upper leaves have shorter stalks, and the narrow, ribbed seed-pods, about 2 inches long, are on very short sturdy stalks. This conspicuous plant often grows in large groups below walls and hedges. (*April—June*).

3 Wild Radish (*Raphanus raphanistrum*). This plant grows about 2 feet tall, and both stem and leaves carry stiff, spreading hairs. The upper leaves are smaller and less deeply lobed than the lower ones. The petals, though usually white with coloured veins, are sometimes yellow. The seed-pods, which end in a long, pointed 'beak', break up into sections each containing 1 seed. (*May—September*).

The rather rare Sea Radish (*R. maritimus*), found on sea cliffs and near beaches, has yellow petals and short-beaked seed-pods which do not easily break into pieces. (*June—August*).

Horse-radish (*Armoracia rusticana*). A tall, cultivated plant with a thick root used for flavouring sauces and large, wavy, Dock-like basal leaves. There are numerous fairly small white flowers and rounded seed pods which rarely ripen in Britain. It has become naturalized in waste places, fields, and by streams throughout Britain, although rarely in Ireland. (*May—July*).

4 Hairy Bitter-cress (*Cardamine hirsuta*). A small annual, about 6 to 12 inches high, with leaves mostly at the base and, in spite of its name, only slightly hairy. Instead of 6 stamens as is usual in this family, the

flowers have 4. The sides of the seed-pods split open and curl upwards to release the seeds (Fig. 1). It is a common plant of rocky and waste places throughout the British Isles. (*March—October*).

Wood Bitter-cress (*C. flexuosa*), common in damp places, has fewer basal leaves and more stem leaves. The flowering stems are somewhat zig-zag shaped, and there are 6 stamens per flower. Large Bitter-cress (*C. amara*), a plant of wet meadows and streamsides, has larger flowers with spreading petals and violet anthers.

Fig. 1

5 Scurvy-grass (*Cochlearia officinalis*). The smooth, fleshy leaves of this little plant contain vitamin C and were at one time eaten by sailors to prevent scurvy. It is fairly common in salt-marshes and on sea-shores round the coasts of Britain. (*May—August*).

Danish Scurvy-grass (*C. danica*), with Ivy-shaped lower leaves and stalked stem leaves, is locally common near the sea. Long-leaved Scurvy-grass (*C. anglica*), found on muddy shores and by estuaries, has the lower leaves tapering to the base and the upper ones clasping the stem. The small, tufted Mountain Scurvy-grass (*C. alpina*), growing on wet rocks and by streams in the mountains, has non-fleshy leaves, white or pinkish flowers, and a pod narrowed at both ends.

6 Sea Rocket (*Cakile maritima*). A fleshy, smooth plant of sandy shores, with much-branched stems growing along the ground or spreading upwards. The upper part of the fruit breaks off when ripe, leaving the small bottom part behind. (*June—August*).

Seakale (*Crambe maritima*). A Cabbage-like plant of sea-shores and cliffs, with a bloom on the large leaves. The flowers are white and sweet-scented. It is often cultivated in gardens because the young shoots are eaten as a vegetable. (*June—August*).

1 WATERCRESS 2 JACK-BY-THE-HEDGE 3 WILD RADISH
4 HAIRY BITTER-CRESS 5 SCURVY-GRASS 6 SEA ROCKET

LIFE SIZE

CRUCIFERAE: CABBAGE FAMILY (See p. 209)

1 Field Penny-cress (*Thlaspi arvense*). The most conspicuous part of this plant is the flat, round, winged fruit, with a notch at the top. The stem grows about a foot high and is hairless, as are the leaves. The basal leaves soon wither. The plant gives off an unpleasant scent when it is bruised. The flowers, like most Cruciferae, have 4 sepals and petals, 4 long and 2 short stamens, and a central ovary. This is a rather uncommon plant of fields and waste places, although sometimes a troublesome weed. (*May—July*).

2 Hairy Rock-cress (*Arabis hirsuta*). A hairy plant, up to 1 or 2 feet tall, usually with an unbranched stem. The stem leaves and short-stalked seed-pods point straight upwards. It is a fairly common plant of dry grassland, hedgebanks, and walls on chalky soils throughout Britain. (*May—August*).

3 Shepherd's Purse (*Capsella bursa-pastoris*). This very common and familiar weed can be recognized by its triangular or heart-shaped seed-pods. The stem grows a few inches to a foot tall from a rosette of leaves that may be deeply cut or almost undivided. The upper leaves clasp the stem. This weed of cultivated land and waste places is known throughout the world. (*January—December*).

4 Thale Cress (*Arabidopsis thaliana*). A plant rather similar to Hairy Rock-cress, but the leaves and seed-pods are not held so upright and the seed-pods are on longer, spreading stalks. It varies in height from a few inches to nearly 2 feet. It is fairly common on walls and dry banks in all parts of the British Isles. (*April—May*).

5 Pepperwort (*Lepidium campestre*). The single stem grows up to 2 feet in height and is sometimes branched. The flowers are very small, and the stamens have yellow anthers. The seed-pods are winged, with a short beak (the style) in the notch at the tip (Fig. 1*a*). Pepperwort is a rather uncommon plant of dry fields and roadsides, not often found in Scotland and Ireland. (*May—August*).

Smith's Cress (*L. smithii*) has several branched stems, growing along the ground at first and then turning upwards. The stamens have violet anthers, and the more oval seed-pods have a longer beak in the notch (Fig. 1*b*). This perennial is widespread and fairly common in fields and by roadsides.

Narrow-leaved Cress (*L. ruderale*) is a branched, hairless, unpleasant-smelling plant with the lower leaves divided into narrow segments and small, narrow upper leaves. The petal-less flowers are minute, and have only 2 stamens. It grows in waste places near the sea in England.

a   x2   b   x2

Fig. 1

Hoary Cress (*Cardaria draba*). A perennial with broader grey-green leaves and flowers in a more or less flat-topped mass. The fruits are very small and unwinged. It is locally common on waste ground and in fields in England and Wales. (*May—July*).

6 Wart-cress or Swine-cress (*Coronopus squamatus*). The plant spreads along the ground, producing flowering shoots opposite the leaves. The fruits are rough and wrinkled, on short stalks. This is fairly common in waste and much trodden places in the south, but rarer in the north. (*May—September*).

Lesser Wart-cress (*C. didymus*), which grows in similar places, usually has spikes of tiny flowers with only 2 stamens, and fruits on longer stalks.

7 Whitlow Grass (*Erophila verna*). The leaves are at the base of the plant, and the leafless flowering stems grow only a few inches high. The flowers have each of the 4 petals divided into 2 lobes. It is a common plant of walls and rocks, almost throughout Britain. (*March—June*).

Shepherd's Cress (*Teesdalia nudicaulis*). This is also a small annual with all the leaves at the base but, unlike the last species, the leaves are divided into leaflets (Fig. 2). Two of the petals are larger than the other two. This plant is locally common on sandy and gravelly ground in England and Wales; rare in Scotland and Ireland. (*April—June*).

x1

Fig. 2

LIFE SIZE

1 FIELD PENNY-CRESS   2 HAIRY ROCK-CRESS   3 SHEPHERD'S PURSE
4 THALE CRESS   5 PEPPERWORT   6 WART-CRESS
7 WHITLOW GRASS

71

CARYOPHYLLACEAE: PINK FAMILY (See p. 210)

1 White Campion (*Melandrium album*). The hairy, and sometimes slightly sticky, flowering stems are up to 3 feet high, and bear soft, hairy leaves. The faintly scented male and female flowers are on separate plants: the male flowers have 10 stamens and the female ones 5 styles. A hairy calyx surrounds the fruiting capsule, which opens by 10 nearly straight teeth. Pink-flowered hybrids between this species and Red Campion (*see* p. 107) are sometimes found. White Campion is common throughout the British Isles in fields, hedgerows, and waste places. (*May—September*).

2 Red Campion, white form (*M. rubrum*). Red Campion, shown on page 107, has a rare white-flowered form, very like White Campion, but with usually smaller, unscented, shorter-stalked flowers, and shorter capsule teeth which roll back when the fruit is ripe.

3 Nottingham Catchfly (*Silene nutans*). A softly hairy perennial up to 1 or 2 feet high, rather sticky towards the top and with often drooping flowers on sticky stalks. The flowers have 3 styles, and the capsule opens by 6 teeth, unlike White Campion. Nottingham Catchfly is a rare plant of dry, stony places and sea cliffs in England and Wales. (*May—July*).

The more widespread annual Small-flowered or English Catchfly (*S. anglica*), growing in sandy fields and waste places, especially in the Channel Isles, is about 1 foot high. The short-stalked, white or pink flowers are smaller and more erect than those of Nottingham Catchfly (Fig. 1). (*June—September*).

4 Bladder Campion (*S. cucubalus*). The hairless, swollen calyx, which gives the plant its name, has a network of veins visible on its outer surface. The plant grows up to 2 feet or higher, and the leaves are usually smooth, with a bloom. The petals are often small and withered looking, and the 6-toothed capsule is hidden within the calyx. This is a common plant of fields and roadsides throughout Britain, especially in the south. (*June—September*).

5 Sea Campion (*S. maritima*). This perennial, woody at the base, has stems spreading outwards and growing along the ground. The leaves are smaller and stiffer than those of Bladder Campion, and there are fewer, long-stalked flowers on each stem. This plant is fairly common on shingly beaches and by mountain streams. (*June—September*).

6 Water Chickweed (*Myosoton aquaticum*). A weak, trailing plant with ribbed stems and hairless, fleshy leaves. The flowers are on hairy stalks which droop after flowering. There are 10 stamens and 5 styles, and a 5-toothed capsule, each tooth being divided in two. Water Chickweed grows in marshes and by streams and ditches in England, Wales, and southern Scotland. (*July—September*).

x1

Fig. 1

1 WHITE CAMPION   2 WHITE form of RED CAMPION   3 NOTTINGHAM CATCHFLY

4 BLADDER CAMPION   5 SEA CAMPION   6 WATER CHICKWEED

LIFE SIZE

CARYOPHYLLACEAE: PINK FAMILY (See p. 210)

1 Bog Stitchwort (*Stellaria alsine*). A slender, hairless perennial, with a weak, squarish stem and all the leaves unstalked. It can be recognized by its 5 petals, which are shorter than the sepals and each almost completely divided into 2 lobes. It is common in wet places throughout the British Isles. (*May—July*).

Marsh Stitchwort (*S. palustris*) is a much less common plant of wet places, and not found at all in the extreme north and parts of Ireland. It has longer, narrower leaves than Bog Stitchwort, and larger flowers, with petals much longer than the sepals. (*May—August*).

2 Greater Stitchwort (*S. holostea*). The hairless, straggly stems grow up to 2 feet long. The unstalked leaves are rough on the edges, and the petals are lobed to about halfway down and are longer than the sepals. The name of Stitchwort is said to have been given because the plants were thought to cure a 'stitch' — a pain in the side of the body. This species is common in woods and hedges throughout Britain. (*April—June*).

Lesser or Heath Stitchwort (*S. graminea*), a common and more slender plant of hedges and heaths, differs from Greater Stitchwort in having smooth-edged leaves and deeply lobed petals. The flowers are smaller and more numerous than those of Marsh Stitchwort. (*May—August*).

3 Common Mouse-ear Chickweed (*Cerastium vulgatum*). A perennial with stems growing along the ground and sending up flowering shoots. The lower leaves are stalked, and both leaves and stems are hairy. The fruits are on fairly long stalks. This grows in waste places and is a very common weed of cultivated land. (*April—September*).

4 Sticky Mouse-ear Chickweed (*C. glomeratum*). The leaves are a lighter green than those of Common Mouse-ear Chickweed, and some of the hairs are tipped by glands. The fruit stalks are shorter, and the flowers rather smaller. It is common in damp places and on waste ground throughout Britain. (*April—September*).

Field Mouse-ear Chickweed (*C. arvense*) has long non-flowering stems creeping along the ground, short and narrow leaves, and some of the hairs tipped by glands. The flowers are large, with petals much longer than the sepals. It grows in dry fields and on banks in most parts of Britain, but is not common. (*April—August*).

Alpine Mouse-ear Chickweed (*C. alpinum*) is smaller and more hairy than Field Mouse-ear Chickweed, with broader leaves. It is found in wet mountainous places. (*June—August*).

5 Dark-green Mouse-ear Chickweed (*C. tetrandrum*). A sticky annual, with spreading branches, often growing along the ground. It differs from most Chickweeds in very often having the parts of its flowers in fours instead of fives. The petals are only slightly divided, and the fruits are on long stalks. It is fairly common in waste places near the sea. (*April—September*).

Little Mouse-ear Chickweed (*C. semidecandrum*) is not unlike Dark-green Mouse-ear Chickweed but the leaves are paler green, the parts of the flowers are in fives, and the stalks of the fruits bend downwards at first. It is common in dry waste places and on bare ground. (*April—June*).

6 Upright Chickweed or Dwarf Chickweed (*Moenchia erecta*). A tiny hairless plant, with a bloom on the leaves. The almost unlobed petals are shorter than the sepals. There are usually 4 stamens and styles, and the capsule splits open by 8 teeth. This Chickweed of dry stony places in England and Wales, especially near the sea, is rather rare. (*April—June*).

7 Chickweed (*Stellaria media*). This very common annual weed has almost hairless leaves and a line of hairs down one side of the rounded stem. The 5 petals are so much divided that they look like 10, and the 3 to 7 stamens have red anthers. Chickweed used to be used to make ointments for rashes and inflammation. It is found throughout the British Isles. (*February—November*).

Lesser Chickweed (*S. apetala*) has small leaves, no petals, and 1 to 3 stamens. It is common on light soils. (*March—May*). Greater Chickweed (*S. neglecta*) is taller, often a perennial, with large leaves and usually 10 stamens. It grows in shady places, mostly in the west. (*April—July*).

1 BOG STITCHWORT   2 GREATER STITCHWORT   3 COMMON MOUSE-EAR CHICKWEED
4 STICKY MOUSE-EAR CHICKWEED   5 DARK-GREEN MOUSE-EAR CHICKWEED   6 UPRIGHT CHICKWEED
7 CHICKWEED

LIFE SIZE

CARYOPHYLLACEAE: PINK FAMILY (See p. 210)

1 Knotted Pearlwort (*Sagina nodosa*). A tufted perennial, about 6 inches high, often with clusters of small leaves in the axils of the stem leaves. The Pearlworts and Sandworts, shown on this page, differ from the Stitchworts and Chickweeds (p. 75) in having flowers with unlobed petals. The petals of Knotted Pearlwort are considerably longer than the sepals, which alternate with the 5 styles. The ripe capsule divides into 5 parts (Fig. 1). This is a fairly common plant of damp, sandy places. (*July—September*).

The smaller Heath Pearlwort (*S. subulata*), which grows on dry heaths, mainly in the north, has clusters of narrow, sharp-tipped leaves, and petals no longer than the sepals. There are pin-shaped hairs on the leaves and flower stalks. (*June—September*).

Fig. 1

2 Sea Sandwort (*Honkenya peploides*). A fleshy plant common on dunes and sandy places near the sea. It has rows of broad leaves and greenish-white flowers. Some plants have male flowers with petals about as long as the sepals; others have female flowers with tiny petals. There are usually 3 styles and a 3-toothed capsule. (*May—August*).

3 Vernal Sandwort (*Minuartia verna*). A small tufted plant, with stiff, narrow leaves and petals slightly longer than the sepals. It differs from the Pearlworts

Fig. 2

in having 3 styles and a 3-toothed capsule. This Sandwort is a rare plant of chalky soils and rocks in the west and north. (*May—September*).

Fine-leaved Sandwort (*M. tenuifolia*) has numerous flowers on slender, spreading stalks, and petals shorter than the sepals (Fig. 2). It is found on walls and dry places, mostly in the east. (*May—June*).

The related Scottish mountain plant, Mossy Cyphal (*Cherleria sedoides*), forms little cushions and has very short-stemmed, greenish flowers, usually without petals. (*June—August*).

4 Corn Spurrey (*Spergula arvensis*). A somewhat sticky annual, with small scaly stipules at the base of the clusters of leaves. The flowers have petals about as long as the sepals, 5 or 10 stamens, and 5 styles. The capsules split open into 5 parts. Corn Spurrey is a common and tiresome weed of cultivated land. (*June—September*).

5 Thyme-leaved Sandwort (*Arenaria serpyllifolia*). A small, rough annual, with unstalked leaves, petals much shorter than the sepals, 3 styles, and a capsule with 6 short teeth at the top only. It is common on walls and dry bare soil throughout Britain. (*April—August*).

6 Three-nerved Sandwort (*Moehringia trinervia*). Very similar to the last species but larger, with stalked leaves that are distinctly 3-veined. The capsule splits into 6 parts down to the base. This is a common annual of woods and damp shady places, throughout most parts of Britain. (*April—July*).

7 Procumbent Pearlwort (*Sagina procumbens*). A perennial with creeping and rooting stems and rosettes of pointed leaves. The flowers are on long stalks that bend down after flowering, but straighten up again when the fruit is ripe. The parts of the flower are in fours, with the stamens opposite the sepals, which are often tinged with red, and the petals very tiny or missing altogether. It is common on paths and waste and grassy places. (*May—September*).

Common Pearlwort (*S. apetala*). A common annual of paths, walls, and dry places throughout Britain, although rare in the far north. Many branches spread outwards from the base, bearing small leaves often fringed with hairs. The flowers are like those of Procumbent Pearlwort, but the flowering stalks remain upright after flowering.

A less common species of sea coasts and mountains is Sea Pearlwort (*S. maritima*), with spreading or tufted stems and small, somewhat fleshy leaves.

1 KNOTTED PEARLWORT   2 SEA SANDWORT   3 VERNAL SANDWORT
4 CORN SPURREY   5 THYME-LEAVED SANDWORT   6 THREE-NERVED SANDWORT
7 PROCUMBENT PEARLWORT

LIFE SIZE

ROSACEAE: ROSE FAMILY (See p. 212)

1 **Raspberry** (*Rubus idaeus*). A woody plant with prickly stems up to 5 feet high and biennial sucker shoots coming from perennial roots. The lower leaves often have 2 pairs of leaflets as well as the end leaflet. The under side of the leaves is white with hairs. The ripe red edible fruits differ from those of Blackberries and other related species in that they come away from the receptacle (swollen top of the flower stalk), leaving a hollow in the fruits, instead of coming off with the receptacle. Raspberry is fairly common in woods and on heaths, especially in the more hilly parts of Britain. (*June—August*).

2 **Burnet Rose** (*Rosa spinosissima*). A small shrub, a foot or more in height, with bristles and straight prickles on the branched stems. The flowers are usually borne singly, without bracts, and the undivided sepals remain upright on top of the purplish-black swollen end of the flower stalk (the 'hip'), which has become enlarged and surrounds the true fruits (achenes). This Rose is fairly common on mountain ledges, dunes, and heaths, especially near the sea, in most parts of the British Isles. (*May—September*).

3 **Field Rose** (*R. arvensis*). A shrub with trailing stems and hooked prickles, which differs from the Roses described on page 116 in having the long styles joined together in the centre of the unscented flowers. The sepals soon drop off and do not appear on the smooth, red hips. This is a fairly common plant of hedges and woods in the south, becoming increasingly rare towards the north. (*June—July*).

A less common southern Rose, *R. stylosa*, has the short styles joined to form a cone-shaped column shorter than the stamens. The white or pink flowers are borne on upright or arched stems.

4 **Blackberry** or Bramble (*Rubus fruticosus*). This is a white-flowered form of the shrub described on page 116.

5 **Dewberry** (*R. caesius*). A plant with weak, creeping stems and rather soft prickles. The leaves, with 3 leaflets, are usually green on both sides. There are only a few flowers, and the fruits are covered with a bluish bloom. It is a common plant of hedges, scrub, and waste places in England and Wales, but less common in Scotland and Ireland. (*May—September*).

Stone Bramble (*R. saxatilis*). The non-flowering, non-woody stems grow along the ground and send up upright flowering stems, about 12 or 18 inches high, with a few, small flowers (Fig. 1*a*). The stems are slightly hairy, and have few prickles or none. The fruits, red when ripe, consist of about 2 to 6 separate rounded fruitlets (Fig. 1*b*). Stone Bramble is a rather uncommon plant of open woods and rocky places in the hills of the west and north. (*June—August*).

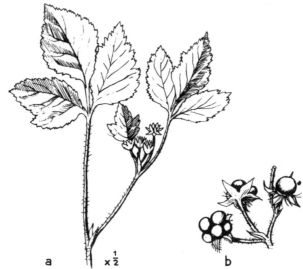

a × ½ b

Fig. 1

78

LIFE SIZE

1 RASPBERRY   2 BURNET ROSE   3 FIELD ROSE
4 BLACKBERRY   5 DEWBERRY

ROSACEAE: ROSE FAMILY (See p. 212)

1 Cloudberry (*Rubus chamaemorus*). A small plant, about 6 inches high, related to the Blackberries (p. 117), but not woody or prickly. The single flower is carried on a long stalk at the end of the stem and is either male (with stamens) or female (with carpels). The ripe fruit is orange. Cloudberry is locally common on moors and in bogs in the north, spreading mainly by creeping underground stems. (*June—August*).

2 Mountain Avens (*Dryas octopetala*). An attractive and distinctive but not common plant. Its regularly-cut glossy leaves are covered with a white felt of hairs underneath. The flowers usually have 8 large petals, and both sepals and flower stalks are sprinkled with dark hairs. The long, hairy styles form feathery plumes on the fruits (achenes) (Fig. 1). This plant of chalk and limestone in N. England, Scotland, and parts of Ireland, is one of a group of plants found both in the arctic and in mountain regions further south. (*June—August*).

Fig. 1

petals instead of 5 (Fig. 2). The flowers are almost unscented. (*May—August*).

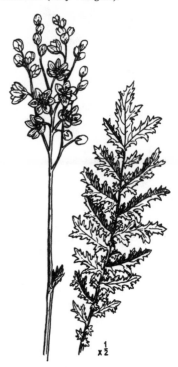

Fig. 2

3 Meadow-sweet (*Filipendula ulmaria*). This plant, with its masses of creamy flowers, might be taken for one of the Parsley family. The flower stalks, however, are not arranged in umbels (*see* p. 46), and there are numerous stamens and carpels. The stems are up to 4 feet tall, with leafy stipules at the bottom of the leaf stalks. The small leaflets between the large ones are characteristic of this and the next species. Meadow-sweet has up to 5 pairs of large leaflets on the lower leaves, which are often white and hairy underneath. The flowers often grow in profusion in damp meadows, fens, and woods, and have a sweet, sickly scent. (*June—September*).

Dropwort (*F. vulgaris*). A similar plant to Meadow-sweet, locally common on dry chalk grassland and in open woods. It has many more pairs of finely cut leaflets on the lowest leaves and often 6 sepals and

4 Wild Strawberry (*Fragaria vesca*). A perennial plant sending out long runners and hairy leaves — similar to the garden Strawberry but smaller, with sweet, juicy 'fruits'. The fleshy part of the 'fruit' is formed from the swollen base of the flower stalk (receptacle) and the true fruits (achenes) are the small, hard 'pips' on the surface. This plant is common in woods and hedges throughout the British Isles. (*April—July*).

5 Barren Strawberry (*Potentilla sterilis*). A usually smaller plant than Wild Strawberry, with bluish-green leaves, and petals with gaps between them, whereas those of Wild Strawberry are almost overlapping. The fruits are hairy achenes on a hairy, not swollen, receptacle. It is common in most parts of Britain on hedgebanks and in woods and scrub, mostly on dry soils. (*February—May*).

LIFE SIZE

1 Cloudberry   2 Mountain Avens   3 Meadow-sweet
4 Wild Strawberry   5 Barren Strawberry

1 **White Clover** or Dutch Clover (*Trifolium repens*, family *Papilionaceae*). A small, hairless perennial, with creeping and rooting stems. The finely-toothed leaflets usually have a pale band across them, as shown in the picture. The flowers turn brown and bend downwards after flowering. This is a very common plant of fields and grassy places throughout the British Isles. (*May—September*).

The hairy Subterranean Trefoil or Burrowing Clover (*T. subterraneum*) is so named because the flower-heads of a few creamy flowers (Fig. 1) bend down to the ground after flowering, and the tiny pods surrounded by swollen calyxes become buried in the earth. It is a rather uncommon plant of dry fields in the south. (*May—June*).

Rough Trefoil or Rough Clover (*T. scabrum*) is a hairy, usually prostrate annual with short leaf stalks and small unstalked white flowers (Fig. 1). It is widespread but local in dry fields, often near the sea. (*May—August*).

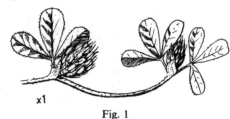

x1

Fig. 1

2 **English Stonecrop** (*Sedum anglicum*, family *Crassulaceae*). A small perennial with spreading stems only a few inches high, and fleshy, hairless leaves. The almost unstalked flowers are often tinged with pink and have 5 sepals, petals, and carpels, and 10 stamens. English Stonecrop grows on rocky places and bare or waste ground, mostly in the west. (*June—August*).

White Stonecrop (*S. album*) is a larger, rare plant of walls and rocks, with longer, oblong, green leaves and more numerous flowers.

3 **White Climbing Fumitory** (*Corydalis claviculata*, family *Fumariaceae*). A slender, smooth plant, climbing by means of the branched tendrils at the ends of the leaf stalks. The flowering stem carries a few, very short-stalked flowers. This Fumitory is a rather uncommon plant of woods and stony places in hilly districts. (*June—September*).

4 **Purging Flax** or Fairy Flax (*Linum catharticum*, family *Linaceae*). A slender annual, with thin, unbranched stems, less then a foot high, and hairless, 1-veined leaves. It differs from other British Flaxes in having opposite leaves and small, white flowers. The parts of the flowers are in fives. Purging Flax is very common on heaths, meadows, and dunes throughout Britain, especially on chalk. As its name indicates, it was at one time used in medicines as a laxative. (*May—October*).

5 **Sundew** (*Drosera rotundifolia*, family *Droseraceae*). This is an insectivorous plant: that is, insects are caught by the sticky, gland-tipped hairs on the spoon-shaped leaves and digested by a juice produced by the leaves, so providing food material for the plant. The flowers on the long, leafless flowering stems nearly all face one way. Sundew is a common bog plant, often growing with Sphagnum moss. (*June—September*).

Great Sundew (*D. anglica*), a less common species, growing in similar places, has more upright, narrow leaves, tapering gradually to longish stalks. The flowering stems may be twice as long as the leaves.

Long-leaved Sundew (*D. intermedia*) grows in slightly drier places and is about mid-way between the two previous species in appearance. The shorter flowering stems come from the side of the plant below the fairly narrow leaves and curve upwards.

6 **Grass of Parnassus** (*Parnassia palustris*, family *Parnassiaceae*). The hairless lower leaves are long-stalked and similar in shape to the unstalked leafy bract about halfway up the flower stalk. The 5 fertile stamens alternate with 5 sterile ones that have split up into threads, with a gland at the end of each. These shiny glands help to attract insects to the flowers. This is a rather uncommon plant of boggy places, mainly in the north. (*July—October*).

7 **All-seed** (*Radiola linoides*, family *Linaceae*). This tiny, much branched plant, though related to the Flaxes, differs from them in having the parts of the flowers in fours. It occurs on damp, sandy or peaty places throughout Britain, but is rather rare. (*June—September*).

1 White Clover   2 English Stonecrop   3 White Climbing Fumitory

4 Purging Flax   5 Sundew   6 Grass of Parnassus

7 All-seed

LIFE SIZE

1 Intermediate Wintergreen (*Pyrola media*, family *Pyrolaceae*). Similar to Common Wintergreen but with darker green leaves and larger flowers, with the straight style protruding beyond the stamens and petals. It is a very local plant of pinewoods in Scotland, and very rare elsewhere in Britain. (*June—August*).

2 Common Wintergreen (*P. minor*). A small perennial with fairly long-stalked, hairless leaves coming from the base, and a leafless flowering stem. The parts of the flowers are in fives, and the petals are not joined together at the bottom. There is a single style, shorter than the stamens and petals and having 5 stigmas. It is a rather rare plant of damp places and woods, especially pinewoods, and found more often in the north than the south. (*June—September*).

Larger Wintergreen (*P. rotundifolia*). The dark green, rounded leaves are on fairly long stalks. The white petals spread out almost flat, and the style is long and curved (Fig. 1). It is a rare, scattered plant of heaths and woods.

Fig. 1   x2

Toothed Wintergreen or Yavering Bells (*Ramischia secunda*). The smallest of the Wintergreens, with pointed tips to the leaves and a number of greenish-white flowers all facing to one side of the flowering stem, with long, straight, protruding styles. This rare Wintergreen grows on heaths and in woods in the north. (*July—August*).

3 One-flowered Wintergreen or St. Olaf's Candlestick (*Moneses uniflora*). Another very rare plant of Scottish pinewoods, distinguished from the other Wintergreens by its solitary flower with a long, thick style. The leaves often appear to be in whorls of 3 or 4 and have short stalks. (*June—August*).

4 Wood-sorrel (*Oxalis acetosella*, family *Oxalidaceae*). A creeping perennial, not more than 6 inches high, with leaf and flower stalks all coming from the rhizome. The soft, hairy leaves tend to fold their leaflets as one is doing in the picture. The white, pale pink, or lilac flowers have 10 stamens and 5 styles; the fruit is a capsule. It is a common woodland plant, often found in deep shade, as in beechwoods. (*April—July*).

Two yellow-flowered species that occur as weeds in gardens in the south are Procumbent Yellow Sorrel (*O. corniculata*), with hairy, often purple leaves on the creeping stems and usually pairs of flowers, and Upright Yellow Sorrel (*O. stricta*), with upright stems and often more than 2 small flowers on a stalk. Both flower later than Wood-sorrel.

5 Meadow Saxifrage (*Saxifraga granulata*, family *Saxifragaceae*). A hairy perennial, with bulbils (swollen buds) at the base and leafy flowering stems, sometimes up to a foot or more high. It is larger than most British Saxifrages and has broad, toothed or lobed leaves. It is locally common on grassland, mainly in the east. (*April—June*).

Rue-leaved Saxifrage (*S. tridactylites*), which grows on walls, rocks, and dry sandy places, is a small hairy annual, with 3- to 5-lobed leaves and tiny flowers on thin stalks (Fig. 2). (*April—June*).

Starry Saxifrage (*S. stellaris*), common in wet places on mountains in the north, has rosettes of unlobed leaves and leafless flowering stems with star-like flowers. The sepals bend downwards after the flowers open. (*June—August*).

Fig. 2

6 Mossy Saxifrage or Dovedale Moss (*S. hypnoides*). A low, tufted, mossy plant with tiny unlobed or 3-lobed, needle-like leaves. The branched flowering stems, up to 6 inches high, are almost leafless. This Saxifrage is locally common on rocks and grassy places in hilly districts in the west and north. (*April—July*).

7 Blinks (*Montia verna*, family *Portulacaceae*). A small hairless plant, with stems sometimes floating or creeping in water, and opposite leaves which may be joined together at the base. The tiny flowers have 2 sepals, 5 petals, 3 stamens, and a style with 3 stigmas. Blinks is common by streams and in fields in the south, but rare in the north. (*April—October*).

1 INTERMEDIATE WINTERGREEN 2 COMMON WINTERGREEN 3 ONE-FLOWERED WINTERGREEN
4 WOOD-SORREL 5 MEADOW SAXIFRAGE 6 MOSSY SAXIFRAGE
7 BLINKS

LIFE SIZE

UMBELLIFERAE: PARSLEY FAMILY (See p. 213)

1 Hemlock (*Conium maculatum*). A very poisonous plant, recognized by the smooth, hollow, purple-spotted stem, and the strong unpleasant smell. The fruits are smooth and rounded, with wavy ribs (Fig. 1*a*). It grows up to 5 or 6 feet high and is common in ditches, damp hedges, and on stream banks, more in the south than the north. Hemlock was used as a poison by the Greeks and Romans, and a sedative drug has been extracted from the leaves and seeds. (*May—August*).

2 Hogweed or Cow Parsnip (*Heracleum sphondylium*). The roughly hairy, thick, grooved, hollow stems growing up to 6 feet tall, and broad, hairy, lobed leaflets are characteristic of this very common plant. The outer flowers have petals of varying sizes, and the fruits are large and flat. Hogweed grows by roadsides and in fields and open woods throughout the British Isles. (*April—October*).

3 Pignut or Earthnut (*Conopodium majus*). A slender, hairless plant, up to 2 feet tall, with a swollen root (tuber) and hollow, ribbed stem. There are few, if any, bracts below the umbels (*see* p. 46), and the petals have long tips curled inwards. The outer flowers often have irregular-sized petals. The fruits are long and smooth. This is a common plant of meadows and woods, except on chalky soil. (*May—June*).

4 Fool's Parsley (*Aethusa cynapium*). This unpleasantly-scented plant is most easily identified by the long bracts, all bent downwards, below each small umbel, on the outer side only. The leaves are like those of the garden Parsley but have a bitter taste — hence the name. The fruits are ribbed and disk-like. Fool's Parsley is a common weed of fields throughout Britain, though rarer in the north. (*July—September*).

Stone Parsley (*Sison amomum*). A hairless hedgerow plant up to 3 feet tall and with a petrol-like smell. The lower leaves are divided into 5 to 9 broad leaflets, while those of the upper leaves are narrow. There are only 3 to 6 small umbels in each compound umbel, and there are bracts at the base of all the umbels. It is common in southern England and Wales. (*July—October*).

Corn Caraway or Corn Parsley (*Petroselinum segetum*). Similar to Stone Parsley, except for the smell, but the lower leaves have more numerous, smaller, dark green leaflets, and the small umbels are on stalks of varying lengths. It is a rather uncommon plant of hedges and damp fields on chalky soil in the south. (*July—September*).

5 Cow Parsley (*Anthriscus sylvestris*). A tall plant, up to 3 or 4 feet high, with hollow, grooved stems and large, fern-like, slightly hairy leaves. The outer flowers in each umbel have petals of different sizes, the largest being on the outside. The bracts below the individual umbels are shorter than those in Fool's Parsley and arranged all round. The long, narrow fruits are smooth (Fig. 1*b*). This plant is very common by roadsides and waste places throughout the British Isles. (*April—June*).

Bur Chervil (*A. neglecta*) is a smaller plant with short-stalked umbels opposite the small leaves, minute flowers, and rough, bur-like fruits (Fig. 1*c*). It is locally fairly common in hedgerows and waste places, especially near the sea.

6 Wild Angelica (*Angelica sylvestris*). The hollow, grooved, purple stems are thick and fleshy, up to 4 or 5 feet high. The large lower leaves often have leaflets arranged in threes. There are usually no bracts below the compound umbels. The white or pinkish flowers have long stamens, and the fruits have 2 broad wings (Fig. 1*d*). Wild Angelica, which is common in damp woods and hedges throughout Britain, is related to the Angelica the stems of which are crystallized and used in confectionery. (*June—September*).

a x 4   b x 4   c x 4   d x 4

Fig. 1

ONE-QUARTER LIFE SIZE

1 HEMLOCK   2 HOGWEED   3. PIGNUT
4 FOOL'S PARSLEY   5 COW PARSLEY   6 WILD ANGELICA

UMBELLIFERAE: PARSLEY FAMILY (See p. 213)

1 Fool's Watercress or Marshwort (*Apium nodiflorum*). A hairless perennial, with stems growing along the ground and upright flowering shoots, 1 to 3 feet high. It differs from other Umbelliferae because the groups of flowers (compound umbels, *see* p. 46) have either very short or no stalks. There are leafy bracts below each stalked individual group (a simple umbel). It is common by streams and in ponds and ditches and is sometimes mistaken for Watercress because of some similarity in the leaves. (*July—September*).

2 Narrow-leaved Water-parsnip (*Berula erecta*). A tall perennial, with hollow, ribbed stems and a carroty smell. The lower leaves have up to 10 pairs of leaflets, which are often cut in a characteristically uneven way. The long-stalked umbels have bracts below both the simple and compound umbels. This species is fairly common in ditches and wet places throughout most of Britain, but rarer in Scotland and Ireland. (*July—September*).

Water Parsnip (*Sium latifolium*). Similar to the previous species, but with fewer, larger, more evenly cut leaflets and larger umbels at the end of the stems only. Water Parsnip is a rather uncommon plant of wet places, absent from the north of Scotland and from most of Ireland. (*July—August*).

3 Water Dropwort (*Oenanthe fistulosa*). A poisonous perennial with swollen roots (tubers) and smooth hollow stems. The lower leaves are fern-like, while the upper ones have only a few narrow segments on long, hollow stalks. There are no bracts below the compound umbels. The flowers sometimes have a pinkish tint. This is fairly common in ditches and wet places, except in the north. (*July—September*).

4 Parsley Water Dropwort (*Oe. lachenalii*). This species differs from Water Dropwort in having a ribbed, nearly solid stem, shorter leaf stalks, and usually bracts below the more spreading compound umbels. It is also fairly common in similar places throughout most of the British Isles. (*June—September*).

5 Hemlock Water Dropwort (*Oe. crocata*). A stout, poisonous perennial with swollen roots (tubers) and thick stems up to 5 feet tall. It differs from Nos. 3 and 4 in having large, very much segmented upper leaves and many more small umbels in each compound group. It is a locally common plant of ditches, marshes, and brackish water. (*June—September*).

Fine-leaved Water Dropwort (*Oe. aquatica*). This tall, spreading, poisonous plant sometimes grows in water, in which case the leaves are cut into very narrow segments. The umbels are opposite the leaves as well as at the ends of the stems, and there are fewer small umbels in each compound group than in Hemlock Water Dropwort. It is fairly common in ditches and wet places.

6 Sanicle (*Sanicula europaea*). The name comes from the Latin *sanus*, healthy, for the plant was once used to heal wounds. Sanicle usually has 5-lobed leaves, mostly from the base of the plant. Its umbels are often in long-stalked groups of 3, and its prickly, bur-like fruits are widely distributed by catching on the coats of animals. Sanicle is common in woods on chalky soil throughout Britain. (*May—September*).

7 Shepherd's Needle or Venus's Comb (*Scandix pecten-veneris*). Both English and Latin names refer to the fruits, which have pointed 'beaks' much longer than the rest of the fruit. The umbels, each consisting of only a few flowers, are often in pairs. This is a small, slightly hairy annual, with grooved stems, and fairly common in fields and waste places throughout Britain, but rarer in the north. (*April—August*).

8 Fine-leaved Marshwort (*Apium inundatum*). This is a small plant of shallow pools and ditches, often growing beneath the water. The leaflets are cut into very narrow segments, especially in the submerged leaves. There are usually only 2 or 3 small umbels together on a stalk opposite a leaf. It is fairly common throughout Britain. (*June—August*).

ONE-THIRD LIFE SIZE

1 FOOL'S WATERCRESS   2 NARROW-LEAVED WATER-PARSNIP   3 WATER DROPWORT

4 PARSLEY WATER DROPWORT   5 HEMLOCK WATER DROPWORT   6 SANICLE

7 SHEPHERD'S NEEDLE   8 FINE-LEAVED MARSHWORT

UMBELLIFERAE: PARSLEY FAMILY (See p. 213)

1 Wild Carrot (*Daucus carota*). This plant is closely related to the cultivated Carrot, and its thick root smells of carrots. It is most easily recognized by the feathery bracts below the compound umbels (*see* p. 46), and by the saucer-like shape of these umbels, especially when they reach the fruiting stage. The fruits are very bristly. Wild Carrot is common in fields and by roadsides throughout Britain. (*June—September*).

The rare Sea Carrot (*D. gingidium*) of southern and western coasts has fleshy leaves with broad segments and a flat or slightly rounded top to the umbels at the fruiting stage.

2 Ground Elder or Goutweed (*Aegopodium podagraria*). A very troublesome garden weed, with long, creeping rhizomes that are extremely difficult to eradicate. It is also locally common in waste places and by roadsides. There are no bracts below the umbels, and the leaflets often have one side longer than the other at the bottom. (*May—August*).

3 Rough Chervil (*Chaerophyllum temulum*). A poisonous, strong-smelling plant up to 3 feet tall, readily identified by the solid, spotted stem, which is swollen at the nodes and covered with rough hairs. There are small bracts below the individual umbels and the long, narrow fruits are ribbed (Fig. 1*a*). This is a common hedge plant, except in the north of Scotland and west of Ireland. (*June—August*).

4 Knotted Hedge-parsley (*Torilis nodosa*). A small trailing plant with hairy stems up to a foot long, and small, very short-stalked umbels opposite the leaves. The fruits are rough and the outer ones bristly (Fig. 1*b*). It grows by roadsides and in fields and waste places, mostly in the south. (*May—August*).

5 Upright Hedge-parsley (*T. japonica*). A tall, slender plant with long-stalked umbels at the ends of the solid, wiry stems and several bracts below each simple umbel of white or pinkish flowers. It is very common by

roadsides and in hedges, coming into flower later than Cow Parsley (p. 87). (*July—September*).

Spreading Hedge-parsley (*T. arvensis*) is usually much smaller than the previous species, with fewer flowers and not more than one bract below each simple umbel. It is fairly common in fields and waste places in England and Wales. (*June—September*).

Sweet Cicely (*Myrrhis odorata*). The large, fern-like leaves are covered with soft hairs, and the hollow stems are also hairy. The umbels of white flowers are at the ends of the stems, and the narrow, ribbed, shiny brown fruits are up to an inch in length (Fig. 1*c*). The plant has a pleasant aromatic scent, rather like aniseed. It is fairly common in the north but rare in the south. (*May—June*).

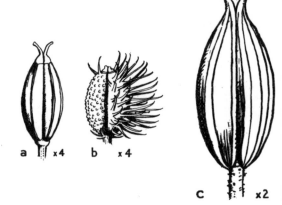

Fig. 1

6 Burnet Saxifrage (*Pimpinella saxifraga*). This slender plant, which is neither a Burnet nor a Saxifrage, can be recognized by the difference in shape between the lower and upper leaves. The slightly hairy stems are 1 or 2 feet high, and there are no bracts below the umbels. It grows in dry fields and by roadsides, especially on chalk soils in the south. (*July—September*).

The less common Greater Burnet Saxifrage (*P. major*) is a larger, stouter plant with hairless, ridged stems, and white, sometimes pinkish, flowers. It grows in shady places in meadows and at the edges of woods. (*June—August*).

1 WILD CARROT

2 GROUND ELDER

3 ROUGH CHERVIL

4 KNOTTED HEDGE-PARSLEY

5 UPRIGHT HEDGE-PARSLEY

6 BURNET SAXIFRAGE

ONE-THIRD LIFE SIZE

1 **Goosegrass** or Cleavers (*Galium aparine*, family *Rubiaceae*). A straggling plant, supporting itself amongst other plants by hooked bristles on the corners of the square stems and on the leaf edges. The leaves are in whorls — a characteristic feature of this family — and the flowers have 4 corolla lobes and 4 stamens. The bur-like fruits catch on to clothes and the fur of animals. Poultry, especially geese, like to eat Goosegrass, which is common in hedges, woods, and bare places throughout Britain. (*May—August*).

2 **Hedge Bedstraw** (*G. mollugo*). A fairly tall, scrambling perennial, up to 3 or 4 feet high, with smooth, square stems and masses of flowers on spreading branches. The one-veined leaves in whorls of 6 to 8 are edged with tiny bristles, but the little wrinkled fruits are smooth. This Bedstraw is common in shady hedges and open woods in the south; less common in the north. (*June—September*).

Northern Bedstraw (*G. boreale*) is a smaller plant, with 3-veined leaves that are usually in fours. The fruits are covered with hooked bristles. It is locally common in mountain pastures and stony places in Wales and the north. (*June—August*).

3 **Woodruff** (*Asperula odorata*). This slender perennial smells of new-mown hay when crushed. It has smooth, squarish, unbranched stems, and whorls of 6 to 8 rough-edged leaves. The corolla tube is longer than in the Bedstraws — about as long as the corolla lobes. The fruit is rough with bristles. Woodruff is common in moist woods on chalky soils. (*April—June*).

4 **Heath Bedstraw** (*Galium hercynicum*). A creeping, slender plant, 6 inches or less high, with smooth, square stems and usually 6 leaves in a whorl. The fruit, though rough, is not bristly. This is a fairly common plant of heaths and fields on chalky soils in the west and north; uncommon elsewhere in Britain. (*June—August*).

Marsh Bedstraw (*G. palustre*). A weak, slightly rough, slender-stemmed plant, with leaves mostly in fours to sixes and with no point at the tip, as most Bedstraws have (Fig. 1a). There are few flowers, and the fruits are smooth. It is common in wet places throughout the British Isles. (*June—August*).

Fen Bedstraw (*G. uliginosum*), a rather smaller, fairly common plant of wet places, differs in having whorls of 6 to 8 leaves with short-pointed tips (Fig. 1b).

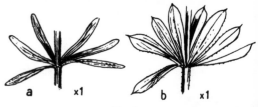

Fig. 1

5 **Squinancy Wort** (*Asperula cynanchica*). A small perennial, a foot or less in height, with smooth, thin stems and usually 4 leaves in a whorl. The small, scented flowers are white or lilac, and the fruit is smooth. It is a locally common plant of chalk grassland and dunes, mainly in the south. (*June—August*).

6 **Marsh Speedwell** (*Veronica scutellata*, family *Scrophulariaceae*). This Speedwell of wet places is smaller than Water Speedwell (p. 175) and has narrower leaves and fewer flowers, which are usually white with pinkish veins, but sometimes blue or pink. There is only one flower spike from each pair of leaves. There are 4 petals and 2 stamens, as in all Speedwells. It is fairly common throughout the British Isles. (*June—August*).

7 **Eyebright** (*Euphrasia officinalis* agg.). A small plant, often no more than 6 inches high, with white hairs on the stem, and unstalked flowers with a 2-lipped corolla characteristically marked with a yellow spot and purple lines on the lower lip. The 4 stamens are under the 2-lobed upper lip. Eyebright is common on heaths and dry fields and has been split up into a large number of species, which are difficult to distinguish. (*June—September*).

8 **Brookweed** (*Samolus valerandi*, family *Primulaceae*). A small, hairless perennial, with slightly fleshy leaves and small bracts about halfway along the flower stalks. The calyx is partly joined to the ovary, and the 5 stamens are attached to the corolla tube, opposite the corolla lobes. It is fairly common in wet places near the sea. (*June—September*).

1 GOOSEGRASS   2 HEDGE BEDSTRAW   3 WOODRUFF

4 HEATH BEDSTRAW   5 SQUINANCY WORT   6 MARSH SPEEDWELL

7 EYEBRIGHT   8 BROOKWEED

LIFE SIZE

93

1 Black Nightshade (*Solanum nigrum*, family *Solanaceae*). A small, non-woody, poisonous plant, with a cone-shaped group of long anthers in the centre of the flower, as in those of the related Woody Nightshade (p. 131) and Potato. The leaves vary from slightly lobed or toothed to quite untoothed. Black Nightshade occurs in waste places and as a garden weed, mostly in the south. (*July—October*).

Thorn-apple (*Datura stramonium*). A highly poisonous plant, not a native of Britain but becoming naturalized and spreading in waste places and cultivated land. It is a tall plant, with fairly large, toothed leaves and solitary, white or purple flowers, not unlike those of Bindweed, but with a longer, more slender corolla tube and a toothed rim to the corolla (Fig. 1). The fruits are large and prickly. (*June—September*).

$x\frac{1}{3}$

Fig. 1

2 Gromwell (*Lithospermum officinale*, family *Boraginaceae*). A branched perennial, about 2 feet tall, which, like most members of the family is covered with stiffish hairs. The 5-toothed calyx persists round the 4 smooth white nutlets. The 5 stamens are within the 5-lobed corolla tube, which has small scales at its mouth. This is a fairly common plant of waste places and hedges in England and Wales, but rare in Scotland and Ireland. (*June—August*).

Corn Gromwell (*L. arvense*) is similar, but it is an annual about half the size, and has somewhat narrower leaves with less obvious side veins. The nutlets are brownish and wrinkled. It grows fairly commonly on cultivated land in England. (*May—July*).

3 Buckbean (*Menyanthes trifoliata*, family *Menyanthaceae*). A plant of shallow water and bogs, with a creeping rhizome and creeping or floating stems. The leaves are divided into 3 leaflets, and there are long white hairs on the inside of the corolla. There are 2 sorts of flowers, corresponding to the pin-eyed and thrum-eyed flowers of the Primrose (p. 26). The fruit is a capsule, splitting into 2 parts when ripe. Buckbean is fairly common throughout Britain. (*May—July*).

4 Enchanter's Nightshade (*Circaea lutetiana*, family *Onagraceae*). A slightly hairy perennial, with stems up to 1 or 2 feet high, and no relation to the other Nightshades. There are 2 pink or white notched petals, 2 stamens opposite 2 crimson sepals, and a 2-lobed stigma. The 2-seeded fruits are covered with hooked hairs. This plant is common in woods and shady places throughout the British Isles. (*June—August*).

Alpine Enchanter's Nightshade (*C. alpina*) is much smaller and, except for the fruits, hairless. The leaves have fewer, larger teeth, and the stigma is rounded. It is locally common in woods and shady places in northern mountains.

Intermediate Enchanter's Nightshade (*C. intermedia*) has characters between those of the other 2 species, and is also found in mountains in the north.

5 Larger Bindweed (*Calystegia sepium*, family *Convolvulaceae*). A perennial, climbing over hedges and bushes by twining its stem in an anticlockwise direction round those of other plants. The flowers, larger than those of Small Bindweed (p. 124), have 2 large bracts covering the 5-lobed calyx. It is common in the south, becoming rarer towards the north. (*June—October*).

Sea Bindweed (*C. soldanella*) has creeping, not climbing, stems, smaller, more rounded leaves, and pink flowers. It is locally common on sandy coasts round most of Britain. (*June—August*).

1 BLACK NIGHTSHADE   2 GROMWELL   3 BUCKBEAN

4 ENCHANTER'S NIGHTSHADE   5 LARGER BINDWEED

LIFE SIZE

95

LABIATAE: DEADNETTLE FAMILY (See p. 216)

1 White Horehound (*Marrubium vulgare*). An aromatic perennial, with square, hairy stems up to 1 or 1½ feet high, and wrinkled, hairy leaves. Each flower has a ribbed calyx with 10 teeth shaped like tiny hooks. The 4 short stamens are hidden inside the corolla, which has a 2-lobed upper lip (Fig. 1). This rather uncommon plant grows in waste places and by roadsides, but is absent from the north. The leaves are used in making medicines for bronchial and digestive complaints. (*June—October*).

x2

Fig. 1

Catmint (*Nepeta cataria*). This rather rare perennial has tiny hairs on the stems and leaves, and a scent of mint. The square stems, which are up to 2 feet high, carry pairs of stalked, opposite, toothed leaves. The branched flower spike carries whorls of flowers at the top of the stem, each flower having a 5-toothed calyx and a 2-lipped corolla, which is white with red spots. The 4 stamens stick out of the corolla tube. It is found on hedgebanks and waste places, mainly on chalk, but not in Scotland. (*July—September*).

2 White Deadnettle (*Lamium album*). The upright, square stems grow 1 or 2 feet high from a creeping rhizome

x1

Fig. 2

and, like the stalked leaves, are slightly hairy. The flowers have calyxes with 5 long teeth. The corolla has an upper lip which hoods the 4 stamens, and a lower lip with small side lobes and a conspicuous bottom lobe, divided into two (Fig. 2). Bees visit the flowers for the nectar stored at the bottom of the corolla tube. White Deadnettle is common and widespread by roadsides and in waste places, except in the north of Scotland and in Ireland. It is in flower for most months of the year. (*March—November*).

3 Gipsywort (*Lycopus europaeus*). A slightly hairy plant, growing up to 3 feet tall, with the lower leaves often deeply cut. The flowers differ from those of other Labiatae in having only 2 stamens and an unlipped corolla with 4 lobes, all nearly equal in size (Fig. 3). Short-tongued insects can reach the nectar in the short, wide corolla tube. This plant of ditches, marshes, and wet places is common nearly everywhere except in Scotland. (*July—September*).

x2

Fig. 3

4 Cut-leaved Self-heal (*Prunella laciniata*). This rare plant is related to the purple-flowered Self-heal (p. 145), but has white flowers and deeply cut or toothed, hairy leaves. It grows on chalk grassland in the south. (*June—August*).

5 Bastard Balm (*Melittis melissophyllum*). A strongly-scented, large-flowered perennial, up to 2 feet tall, with white hairs on the stems, leaf stalks, and lower surface of the wrinkled leaves. The flowers have irregular 2-lipped calyxes, and the anthers of the 4 stamens lie beneath the upper lip of the slightly-hooded corolla. The flowers are visited by bees and hawk-moths. This is a rare, local plant of woods and hedges in southern England and Wales. (*May—July*).

LIFE SIZE

1 WHITE HOREHOUND   2 WHITE DEADNETTLE   3 GIPSYWORT

4 CUT-LEAVED SELF-HEAL   5 BASTARD BALM

COMPOSITAE: DAISY FAMILY (See p. 218)

1 Yarrow or Milfoil (*Achillea millefolium*). A perennial with creeping underground stems and tough, ribbed flowering stems, 1 foot or more in height. Milfoil means 'a thousand leaves' and refers to the tiny segments into which these are divided. The white or pinkish flowers have a strong, aromatic smell. Each 'flower' consists of about 5 outer (ray) florets, resembling the petals of a single flower, and tiny inner (disk) florets. Yarrow, which is very common by roadsides and in meadows throughout Britain, has the property of stopping bleeding, and used to be used for this purpose, as a tonic, and for colds. (*June—October*).

2 Ox-eye Daisy or Marguerite (*Chrysanthemum leucanthemum*). This large and attractive perennial Daisy has flowering stems up to 2 feet high which spring from rosettes of rounded and toothed leaves on very long stalks. There are several rows of bracts with reddish-brown edges forming an involucre below each flower-head. The receptacle (the enlarged end of the flower stalk bearing the florets) is flat. Ox-eye Daisy is common by roadsides and in meadows and cultivated fields throughout Britain. (*June—August*).

3 Daisy (*Bellis perennis*). A dwarf perennial, with a rosette of leaves and hairy flower stalks a few inches long. 'Daisy' comes from 'Day's eye', and is a name given to several Composites with yellow disk florets and white ray florets. This Daisy is a very common weed of lawns and grows in grassy places throughout the British Isles, flowering practically the whole year. (*March—October*).

4 Sneezewort (*Achillea ptarmica*). A perennial related to Yarrow, but distinguished by its undivided leaves with tiny teeth, and its fewer, larger flower-heads, with 8 or more outer (ray) florets. The bracts forming the involucre are woolly. Sneezewort grows up to 2 feet tall and is fairly common in damp meadows and shady places. (*July—September*).

5 Feverfew (*Chrysanthemum parthenium*). The English name comes from the Latin *febrifugia*, 'driving away fevers', and the plant was once used in medicines for bringing down high temperatures in feverish patients. It is a perennial with an aromatic scent and branched, ribbed stems up to 2 feet high. The lower leaves are stalked and more deeply divided than the upper ones. The involucral bracts are hairy, with deeply cut tips. It is a fairly common wayside plant, growing on walls,

banks, and waste places, but is rare in Ireland. *July—September*).

6 Corn Chamomile (*Anthemis arvensis*). A small, almost scentless annual with softly hairy stems and leaves. About 3 rows of hairy bracts form an involucre round each flower-head. The receptacle is cone-shaped, instead of flat, as with Ox-eye Daisy, and differs from that of Scentless Mayweed in having narrow, pointed scales among the florets (Fig. 1a). This is a fairly common plant in fields and waste places on chalky soil. (*June—September*).

a   x2   b   x2

Fig. 1

A sweet-scented perennial, Chamomile (*A.nobilis*), was once used to make chamomile tea for digestive troubles. It is rather shorter than Corn Chamomile, and has broader and more rounded scales on the receptacle (Fig. 1b). It grows locally on gravelly or sandy places in England and Wales.

Stinking Mayweed (*A. cotula*) has a strong, unpleasant smell. It is less hairy than the previous species, and the outer (ray) florets lack styles (Fig. 2). It is common in fields and waste places in the south, though rarer in the north.

x2

Fig. 2

7 Scentless Mayweed or Corn Feverfew (*Matricaria inodora*). An almost unscented perennial with hairless leaves and smooth spreading stems up to 2 feet long. The receptacle is rounded and solid, and, unlike Corn Chamomile, there are no scales among the florets. The bracts round the flower-heads have brown edges. This common plant of waste land and fields is found throughout Britain. (*July—September*).

Wild Chamomile or Scented Mayweed (*M. chamomilla*) is a closely related though less common annual species with an aromatic scent. It differs from Scentless Mayweed in having involucral bracts with pale edges, and a hollow receptacle like that of Rayless Mayweed (see p. 38).

1 Yarrow   2 Ox-eye Daisy   3 Daisy
4 Sneezewort   5 Feverfew   6 Corn Chamomile
7 Scentless Mayweed

LIFE SIZE

ORCHIDACEAE: ORCHID FAMILY (See p. 219)

1 **Creeping Lady's Tresses** (*Goodyera repens*). The upright shoots from 4 to 9 inches high grow from a creeping stem. The leaves, of which the lowest are short-stalked, have a network of veins. The unstalked, scented flowers are borne in a spike on one side of the stem. There is no spur to the flower, and the unlobed lip has a grooved end which is bent under at the tip. Creeping Lady's Tresses grows in pinewoods in Scotland, and in scattered places in north and east England. (*June—August*).

2 **White Helleborine** (*Cephalanthera damasonium*). The stem is from 6 to 18 inches in height, with brown scales at the base. The leaves are broad, and the bracts below the lower flowers are very long. The twisted ovary is visible below the erect, unspurred perianth segments, which never open very widely. The orange lip is divided into two parts, with a narrow ridge between. This rather rare Orchid grows in woods, especially beechwoods, on chalky soil, mainly in the south. (*May—June*).

Narrow Helleborine or Long-leaved Helleborine (*C. longifolia*). A taller and rarer plant than White Helleborine, with longer, narrower leaves, and with the bracts below the lower flowers shorter than the flowers (Fig. 1). It is found in scattered places in hilly woods, mainly on chalk. (*May—July*).

Fig. 1

$\times \frac{1}{2}$

3 **Butterfly Orchid** (*Platanthera chlorantha*). The flowers of this Orchid can be recognized by the very long spur, longer than the ovary, and by the long, narrow, unlobed lip (Fig. 2*a*). The flower spike is a foot or slightly more in height and springs from 2 large leaves near the base. The fragrant flowers are pollinated by moths. It is fairly common in woods and grassy places throughout Britain, mainly on chalky soils. (*May—August*).

The Lesser Butterfly Orchid (*P. bifolia*) is very similar, but smaller and more slender, with whiter flowers (Fig. 2*b*). It grows in the same kind of places as Butterfly Orchid, but is rather less common.

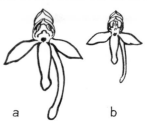

a   b

Fig. 2

4 **Autumn Lady's Tresses** (*Spiranthes spiralis*). A small Orchid, with flowers on one side of a twisted spike, usually about 4 to 6 inches high. It is somewhat like Creeping Lady's Tresses, but flowers later and has no creeping stem. The unstalked leaves are all at the base of the plant and are produced at the same time as the flowers or slightly later. There are a few scale-like leaves up the flower-stem. The almond-scented flowers are unspurred and have an unlobed, greenish lip with a frilly edge. This is a rather rare plant of downs and pastures, usually on chalk, and mainly in the west. (*August—September*).

5 **Heath Spotted Orchis** (*Orchis ericetorum*). This is like a white-flowered form of the Common Spotted Orchid (*O. fuchsii*) described on page 158. It is usually smaller — about 6 to 12 inches high — and the middle lobe of the lip is usually smaller and shorter than the side lobes. It is found commonly on damp, acid soils in the north-west and south-east.

Small White Orchid (*Leucorchis albida*). A small Orchid, one foot or less in height, with a spike of white, scented flowers closely crowded together. The perianth segments form a hood over the 3-lobed lip, and the flower has a short, stout spur. This Orchid is locally common on hilly pastures, mostly in the north, but rare elsewhere. (*May—July*).

1 CREEPING LADY'S TRESSES   2 WHITE HELLEBORINE   3 BUTTERFLY ORCHID
4 AUTUMN LADY'S TRESSES   5 HEATH SPOTTED ORCHIS

LIFE SIZE

1 **Water Soldier** (*Stratiotes aloides*, family *Hydrocharitaceae*). A water plant, appearing above the surface at flowering time. The spiny-toothed leaves, up to 18 inches long, grow in rosettes at the bottom of still water in ponds, ditches, and in the Broads. Male and female flowers are on separate plants. The male flowers, which are the rarer, may have several heads on one stem, while the female are solitary on a short stalk. The plant, which reproduces by rooted runners, the fruits never maturing in England, is locally common in eastern England, but rare elsewhere. (*June—August*).

2 **Frogbit** (*Hydrocharis morsus-ranae*). A floating water plant, with creeping stems from which groups of leaves arise at the nodes. Special buds are formed in the autumn, by which the plant survives the winter. The male and female flowers are separate, the male being several to a stem and the female solitary. The male flowers have 3 narrow sepals, 3 thin petals, and 12 stamens; the female flowers, as well as sepals and petals, have 6 sterile stamens and an ovary with 6 styles (Fig. 1). Frogbit is locally common in ponds and ditches in chalky places, mainly in the south. (*July—August*).

Fig. 1

(*a*) Male flower. (*b*) Male. (*c*) Female.

3 **Ramsons** (*Allium ursinum*, family *Liliaceae*). There are usually 2 shiny leaves with parallel veins, and a triangular flowering stem about 1 foot high. The long bracts at the base of the group of flowers soon wither and fall off. The flowers have 6 petal-like segments, 6 stamens, and a 3-lobed ovary with a single style. The plant, which smells of onions, lasts through the winter as a bulb, and is common in damp and shady places throughout Britain. (*April—June*).

4 **Arrowhead** (*Sagittaria sagittifolia*, family *Alismataceae*). This water plant has creeping stems and three kinds of leaves: thin, narrow submerged leaves, broader floating leaves, and the characteristic arrow-shaped leaves borne on long stalks above the water. The flowering stem rises above the leaves, up to 3 feet long. The bottom flowers are female with numerous carpels and are smaller than the upper male flowers, which have numerous stamens. This is a handsome, fairly common plant of ponds and canals in the south, but is rare in the north. (*July—September*).

5 **Sea Arrowgrass** (*Triglochin maritima*, family *Juncaginaceae*). A marsh plant with a short rhizome and fleshy leaves. The Plantain-like flowering stem is up to 18 inches high. The individual flower stalks carry flowers consisting of 6 tiny, purple-edged perianth segments, 6 stamens, and 6 carpels. The plant is fairly common in salt-marshes and on sea-shores. (*June—September*).

Marsh Arrowgrass (*T. palustris*) is less conspicuous and has more slender leaves than Sea Arrowgrass. There are only 3 fertile carpels. It is a fairly common marsh plant throughout Britain, although less common in the south. (*June—August*).

Common Duckweed (*Lemna minor*, family *Lemnaceae*). This very tiny floating plant often grows in large masses, covering the surface of still water in ponds and ditches. It consists of a small, light green, rounded, flattish, leaf-like structure, called the thallus, and a single root (Fig. 2). The flowers, consisting of stamens and ovary only, are rare and inconspicuous. This is the commonest of several species of Duckweed, and is found throughout the British Isles. (*June—August*).

Fig. 2

1 WATER SOLDIER   2 FROGBIT   3 RAMSONS

4 ARROWHEAD   5 SEA ARROWGRASS

LIFE SIZE

1 Scarlet Pimpernel (*Anagallis arvensis*, family *Primulaceae*). A small, slender plant, with squarish stems often spreading along the ground for about a foot. Black dots (glands) are scattered over the stems and the under side of the leaves. The red (rarely white or blue) flowers have 5 stamens surrounding the central style. The fruit is a capsule, which splits round the middle when it is ripe, the lid, with the style still attached, looking like a little cap (Fig. 1). A country name for this plant is Poor Man's Weather-glass, because the flowers close in bad weather. It is a very common weed of cornfields, gardens, and roadsides throughout the British Isles, except the far north. (*May—September*).

There is a rare, blue-flowered species (*A. foemina*), which has narrower petals and shorter flower stalks. It is found mainly in the south-west.

Fig. 1

2 Hound's-tongue (*Cynoglossum officinale*, family *Boraginaceae*). This biennial, up to 2 or 3 feet high, has a mousy smell, and is covered with long, silky hairs, which make it softer to touch than most members of this family. The lowest leaves are very large and stalked; the upper, smaller ones are in a spiral round the stem and unstalked. The flowers have 5 short stamens attached to the corolla tube, at the mouth of which is a bump on each corolla-lobe. The fruit consists of 4 nutlets covered with prickly hooks. This plant is widespread on dry soils in waste places, at the edges of woods, and on sand-dunes and downs by the sea, but is rather local. (*June—August*).

There is a rare plant of woods and shady places in the south and east, Green Hound's-tongue (*C. germanicum*), which has rougher, greener leaves and bluish-purple flowers. (*May—July*).

3 Pheasant's Eye (*Adonis annua*, family *Ranunculaceae*). The Latin name of this rare plant of cornfields in the south is that of a beautiful youth who was loved by the goddess Venus and who, when he died, was turned into a flower. Pheasant's Eye grows from 6 to 18 inches high, and has smooth, branched stems and solitary flowers. There are 5 green or reddish sepals and 5 to 8 petals. The numerous fruiting carpels are like those of a Buttercup. (*June—August*).

4 Field Poppy or Corn Poppy (*Papaver rhoeas*, family *Papaveraceae*). A plant containing milky juice which

is poisonous to animals. It grows up to 1 or 2 feet high, with spreading hairs on the leaves and flower stalks. The 2 hairy sepals covering the bud drop off as the flower opens. The dark ridges on top of the ovary are the stigmas, still to be seen on the ripe fruit (capsule), which opens by a ring of little holes just below the top (Fig. 2). This Poppy is a common weed in cornfields throughout Britain, except the far north, but is now less conspicuous in many places owing to the increased use of weed-killers. (*June—August*).

A similar but, except in the north, less common species, the Long-headed Poppy (*P. dubium*), has slightly smaller flowers and long, narrow, hairless capsules.

Fig. 2

Long Prickly-headed Poppy or Pale Poppy (*P. argemone*). A small, stiffly hairy Poppy, up to 12 or 18 inches high, with small, pale-red flowers, with a dark blob at the base of each of the 4 petals. The stamens are reddish, and the bristly, ribbed capsules are long and narrow, with usually 5 stigmas (Fig. 3). It is quite common in cornfields in the south; rare in the north. (*May—July*).

The rare Round Prickly-headed Poppy (*P. hybridum*) has deep crimson flowers and egg-shaped, very bristly capsules. It grows mainly on chalk or limestone.

Fig. 3

5 Orange Hawkweed or Fox and Cubs (*Hieracium aurantiacum*, family *Compositae*). The general features of Hawkweeds are described on page 36. This plant, originally an escape from gardens, has naturalized on railway banks and in woods, especially in the north. The stem has short runners, and the green, hairy leaves are mostly in a rosette at the base of the stem. (*June—August*).

LIFE SIZE

1 Scarlet Pimpernel 2 Hound's-tongue 3 Pheasant's Eye
4 Field Poppy 5 Orange Hawkweed

CARYOPHYLLACEAE: PINK FAMILY (See p. 210)

1 **Red Campion** (*Melandrium rubrum*). This plant produces short creeping stems and upright flowering ones from 1 to 3 feet high. Stem and leaves are covered with soft hairs. The unscented male and female flowers are on different plants; the male flowers have 10 stamens and the female ones 5 styles. Pink-flowered hybrids between Red Campion and White Campion (p. 73) are fairly common. The fruit, which is a capsule with 10 teeth, is enclosed by the enlarged calyx. The seeds are black and rough. Red Campion is common in woods and hedges, and on rocky slopes and cliffs in most parts of Britain. (*April—August*).

Night Catchfly or Night-flowering Campion (*M. noctiflorum*). A slightly smaller species than Red Campion, with narrower leaves and stems covered with sticky hairs. The pink or white petals are curled inwards in the daytime, opening in the evening, when the flowers are sweetly scented. It is a rather rare plant of cornfields, found more often on sandy soils. (*July—September*).

2 **Corn Cockle** (*Agrostemma githago*). An annual, from 1 to 3 feet tall, with long silky hairs closely pressed against the stem and leaves. The scentless flowers are visited by butterflies. The plant was once common in cornfields, but has been almost eradicated because the black seeds, being poisonous, make flour containing them unfit for use. Corn Cockle is now a rare weed. (*June—September*).

3 **Red Catchfly** (*Viscaria vulgaris*). A tufted perennial with flowering shoots 1 to 2 feet high. The stems are sticky just below the nodes. The petals are only slightly notched. This is a very rare plant of cliffs and rocks in North Wales and Scotland. (*June—August*).

The even more rare Red Alpine Catchfly (*V. alpina*), found in high places in the Lake District and Scotland, is smaller, not sticky, and has deeply notched pink petals.

4 **Soapwort** or Bouncing Bett (*Saponaria officinalis*). A perennial producing flowering stems 1 to 3 feet high from a creeping rhizome. The leaves are opposite, strongly ribbed, and not hairy. The scented flowers, which have only 2 styles, are visited by hawk-moths. It is fairly common near villages and by streamsides, and the leaves were at one time used in making soap. (*July—September*).

5 **Ragged Robin** (*Lychnis flos-cuculi*). A slightly hairy, rather rough perennial, about 1 to 2 feet tall. It is easily recognized by the untidy-looking divided petals. The flowers are visited by butterflies and bees. It is a fairly common plant of damp places throughout the British Isles. (*May—July*).

Deptford Pink (*Dianthus armeria*). A rare annual of hedges and dry pastures, especially on sandy soil. It grows 1 to 2 feet high, with stiff, narrow, dark green leaves and clusters of bright pink flowers (Fig. 1*a*). It is a close relative of the garden Sweet William. (*July—August*).

Maiden Pink (*D. deltoides*) is shorter, with smaller, greyer leaves. The flowers are borne singly or in pairs and have toothed petals which are pink or white, with spots (Fig. 1*b*). It is a local plant of dry pastures and banks, and is not easy to see in dull weather as the petals close.

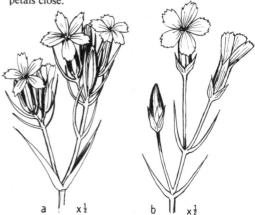

a   x½   b   x½

Fig. 1

Moss Campion (*Silene acaulis*). A tiny, moss-like plant of mountains in the north, with tufts a few inches high (Fig. 2). The leaves are narrow and furrowed, and the purple-pink flowers have 3 styles. (*July—August*).

Fig. 2

1 RED CAMPION   2 CORN COCKLE   3 RED CATCHFLY
4 SOAPWORT   5 RAGGED ROBIN

LIFE SIZE

1 **Policeman's Helmet** or Himalayan Balsam (*Impatiens glandulifera*, family *Balsaminaceae*). This tall Himalayan plant, grown in gardens, has now naturalized by streams and rivers and on waste places, especially in the north and west of England and in Wales. The ribbed, hollow stems bear leaves in opposite pairs or in threes, with a pair of stalked glands at the base of each leaf stalk. There are 3 coloured sepals: 2 small ones at the sides and a large one forming a pouch at the bottom of the flower, with a tiny, hooked spur. The upper petal is large and erect, the others are joined to make the lip. The 5 stout stamens are gathered together round the ovary. When the ripe fruit (a capsule) is touched, it immediately opens to release the seeds — hence another name, Jumping Jack. (*June—September*).

2 **Water-plantain** (*Alisma plantago-aquatica*, family *Alismataceae*). This plant is related to Arrowhead (p. 103), but can be distinguished by the smaller flowers and Plantain-like leaves. Water-plantain is rooted in the mud at the bottom of still or slow-moving water, with the long-stalked leaves mostly appearing above the surface. The flowering stem grows

Fig. 1

up to 3 feet high, bearing numerous 3-petalled flowers, with narrow sepals. The flowers open only in the afternoons. There are 6 stamens and a ring of carpels. This is a fairly common plant of shallow ponds, ditches, and slow-moving rivers, throughout the British Isles, except the far north. (*June—August*).

A similar but less common species, *A. lanceolatum*, has narrower leaves and smaller, pinker flowers, with more rounded sepals and pointed petals, which open mainly in the mornings. It occurs in the south and east.

Lesser Water-plantain (*Baldellia ranunculoides*). A locally common plant of streamsides and marshy places, about 9 inches high (Fig. 1). It has erect, leafless stems, with the pink or purplish flowers in an umbel at the end of the stem, and also often in a whorl lower down the stem. The flowers have 3 green sepals and 3 petals. The leaves are very narrow and on stout stalks. (*May—August*).

3 **Water Violet** (*Hottonia palustris*, family *Primulaceae*). This species is no relation to the other Violets. The leaves are beneath the water surface, and the leafless flowering stem, a foot or slightly more in height, bears whorls of pale pinkish or lilac flowers. The parts of the flowers are in fives. The fruit is a capsule, which opens by 5 slits to release the seeds (Fig. 2). The plant also spreads by branches that break off and float away from the parent. It is locally common in ponds and ditches in England and Wales. (*May—June*).

×2

Fig. 2

4 **Flowering Rush** (*Butomus umbellatus*, family *Butomaceae*). A tall, very handsome water plant, with long, 3-sided leaves and a leafless flowering stem, sometimes 4 feet or more tall, with a single umbel of flowers at the end. There are 9 stamens, 6 opening before the other 3, and there are usually 6 carpels. This is a rather uncommon plant of ponds and water-channels in England and Ireland. (*July—September*).

LIFE SIZE

1 Policeman's Helmet   2 Water-plantain   3 Water Violet

4 Flowering Rush

1 Broad-leaved Willow-herb (*Epilobium montanum*). A perennial with a slightly hairy, rounded stem, not more than 2 feet tall. The sharply toothed leaves are rounded at the base and are usually opposite. The flowers have 4 petals, 4 short and 4 long stamens, and a 4-lobed stigma. The fruit is a long capsule that splits into 4 parts when ripe, setting free the plumed seeds, which are dispersed by the wind. This Willow-herb is common in woods, hedgebanks, and waste places, and as a garden weed, throughout the British Isles. (*June—August*).

A less common species, Spear-leaved Willow-herb (*E. lanceolatum*), differs from the last species in having leaves that taper gradually at the base to a short stalk. The upper leaves are alternate, and the flowers small and pink. It is found in south and south-west England by roadsides and on dry waste places. (*July—September*).

2 New Zealand Willow-herb (*E. pedunculare*). A small perennial with creeping and rooting stems. The single flowers spring from the leaf axils, and the capsules are carried on stalks up to 2 inches long. This species, originally a native of New Zealand, has now naturalized in northern Britain, where it grows on moist rocks and beside streams. (*June—July*).

Alpine Willow-herb (*E. anagallidifolium*). A rather uncommon dwarf plant of mountain streamsides in Scotland and the north of England. The stem, which is less than 6 inches high, has pairs of tiny leaves and 1 to 3 pink flowers (Fig. 1). (*July—August*).

Fig. 1

Chickweed Willow-herb (*E. alsinifolium*), also uncommon, grows in similar places but is slightly bigger, with 2 to 5 bluish-red flowers.

3 Rosebay Willow-herb or Fireweed (*Chamaenerion angustifolium*). A tall, handsome perennial, up to 5 feet high. The long leaves, which often have wavy edges, are arranged in a spiral round the stem. The long capsules split to release numerous tiny plumed seeds, by means of which the plant quickly spreads over large distances. It is very conspicuous in clearings in woods, or on waste land, and was common on bomb-sites after the Second World War. (*July—September*).

4 Marsh Willow-herb (*Epilobium palustre*). A slender perennial with rounded, somewhat hairy stems and narrow, mostly opposite leaves. The drooping, pink or whitish flowers have knob-shaped stigmas, and ovaries covered with short hairs. This is a locally common plant of wet places, but not on chalky soil. (*June—August*).

Square-stemmed Willow-herb (*E. adnatum*). Stouter than the preceding species and distinguished from it by the shiny upper surface of the leaves and the 4 distinct raised lines on the tough stems. The flowers have pale lilac petals. It is a plant of damp woods and wet places, locally common in the south, rare in the north and in Ireland. (*July—August*).

A species of wet places not easy to distinguish from Square-stemmed Willow-herb is Short-fruited Willow-herb (*E. obscurum*). As well as the shorter capsules, the leaves are broader and rounder at the base. The flowers are usually a deeper pink.

5 Great Hairy Willow-herb or Codlins-and-cream (*E hirsutum*). A tall, stout, very hairy perennial, up to 6 feet high. The leaves are all opposite and clasp the stem at the base. The stamens usually shed their pollen before the 4-lobed stigma is fully developed, so the flowers depend on cross-pollination by bees and hoverflies. It is common by streamsides and in damp places throughout most of Britain. (*July—August*).

The Small-flowered Willow-herb (*E. parviflorum*) grows in similar places, but is shorter and more slender, with smaller flowers, and the leaves mostly alternate and not clasping the stem.

HALF LIFE SIZE

1 Broad-leaved Willow-herb   2 New Zealand Willow-herb   3 Rosebay Willow-herb

4 Marsh Willow-herb   5 Great Hairy Willow-herb

1 Lady's Smock or Cuckoo Flower (*Cardamine pratensis*, family *Cruciferae*). A perennial, with smooth stems about 12 to 18 inches high and a basal rosette of stalked leaves, divided into slightly toothed, rounded leaflets. As in all *Cruciferae*, the flowers have 4 sepals and petals and 6 stamens: 4 long and 2 short. The long, narrow fruit is divided down the middle into 2 parts, each with a single row of seeds. Lady's Smock is common in damp meadows and by streams throughout the British Isles. (*April—June*).

2 Pink Purslane (*Claytonia alsinoides*, family *Portulacaceae*). A smooth, fleshy annual up to about 1 foot high, each flower stem bearing a pair of stem leaves. The pink or white flowers have 5 stamens opposite the 5 petals, which are not joined together at the bottom, and a style with 3 stigmas. The fruit is a capsule that splits into 3 parts when it is ripe. This is a rather uncommon plant of damp woods and shady places in the west and north. (*April—July*).

A locally common, small, white-flowered species of waste places, Spring Beauty (*C. perfoliata*), has the stem leaves joined together so that the stem seems to go through the middle of a leaf (Fig. 1).

x1

Fig. 1

3 Common Milkwort (*Polygala vulgaris*, family *Polygalaceae*). The Milkworts, which are most often blue, are shown on page 169. The Common Milkwort, however, is sometimes pink, as shown in this picture.

4 Bird's Eye Primrose (*Primula farinosa*, family *Primulaceae*). A small perennial, growing in mountain pastures in the north. The under side of the leaves is covered with a whitish or yellow mealy substance. The flowers are pin-eyed and thrum-eyed, as in the Primrose (*see* p. 26). (*May—July*).

A very similar but smaller species, *P. scotica*, is found only in the north of Scotland. It has purplish flowers and corolla lobes shorter than the corolla tube.

5 Sea Spurrey (*Spergularia marginata*, family *Caryophyllaceae*). A perennial, with stems spreading outwards for up to one foot. The fleshy leaves have triangular stipules at the base. The pink or white petals are longer than the sepals and there are 10 stamens. This species is locally fairly common in salt-marshes. (*June—September*).

Another salt-marsh species, *S. salina*, has slightly smaller flowers, with petals shorter than the sepals, and fewer than 10 stamens. The Cliff Sand Spurrey (*S. rupicola*), found on rocky coasts in the south and west, has glandular hairy stems, silvery stipules, and petals about the same length as the sepals. Sand Spurrey (*S. rubra*), common in sandy and gravelly places, is a rather hairy plant, with the leaves not fleshy, and the petals of the tiny flowers shorter than the sepals. (*May—September*).

6 Hairy Stonecrop (*Sedum villosum*, family *Crassulaceae*). A small Stonecrop, only a few inches high, with glandular hairs on both the short non-flowering shoots and the longer flowering ones. The unstalked, untoothed stem leaves are arranged alternately. The petals are not joined together at the bottom. This Stonecrop is locally common in damp, rocky places in the northern mountains. (*June—July*).

7 Sea Heath (*Frankenia laevis*, family *Frankeniaceae*). A small, Heather-like, woody perennial, with creeping and rooting stems, up to 6 or 12 inches long. The leaves are often crowded together on very short shoots so that they appear to grow in bunches. The calyxes of the unstalked flowers have 5 teeth and the 5 (rarely 4) petals form a tube at the base. This is a rather rare plant of salt-marshes in south and east England. (*July—August*).

8 Water Purslane (*Peplis portula*, family *Lythraceae*). A small, hairless annual, with square, creeping and rooting stems. The leaves are rather thick and not more than half an inch long. The minute flowers are borne in the leaf axils. There is usually a 6-lobed calyx, 6 pink petals (sometimes none), and 6 or 12 stamens. Water Purslane is fairly common in damp places, especially on bare ground, except in Scotland. (*June—September*).

LIFE SIZE

1 Lady's Smock	2 Pink Purslane	3 Common Milkwort
4 Bird's-eye Primrose	5 Sea Spurrey	6 Hairy Stonecrop
7 Sea Heath	8 Water Purslane	

113

PAPILIONACEAE: PEA FAMILY (See p. 211)

1 Red Clover (*Trifolium pratense*). A rather straggling, hairy plant, with stems up to 2 feet long, and leaflets with a whitish patch on them. The stipules at the base of the leaf stalk are large, ending in long points. There are 2 leaves immediately below the very hairy calyx enclosing each flower-head. Bees visit the sweetly scented flowers, thus cross-pollinating the flowers and producing an excellent honey from the nectar. Red Clover is cultivated in Britain, and also grows wild in grassy places. (*May—October*).

2 Zigzag Clover (*T. medium*). This plant is distinguished from Red Clover by its longer leaflets, with only a faint whitish patch, if any. Only the teeth of the calyx are hairy. The name 'Zigzag' refers to the shape of the straggling stems. The plant grows by roadsides and in grassy places throughout Britain, more frequently in the north. (*June—September*).

Crimson Clover (*T. incarnatum*). This species, first introduced as a crop plant, has now become naturalized in southern Britain. It has broader leaflets than Red Clover and longer crimson flower-heads. The calyx teeth are long and bristle-like.

3 Hare's-foot (*T. arvense*). A slender annual Clover, easily recognized by the soft, long-stalked flower-heads, in which the petals are almost hidden by the hairy calyx teeth. It is locally common in dry, sandy fields, especially near the sea. (*June—September*).

Knotted Clover or Soft Trefoil (*T. striatum*). A low, tufted, annual plant, with long, soft hairs covering the

$x\frac{1}{2}$

Fig. 1

stems and leaves. The small, pink flowers are in stalkless clusters at the ends of the stems and in the leaf axils, partly hidden by the stipules (Fig. 1). Knotted Clover grows in dry sandy fields in most parts of Britain, but only by the coast in Ireland. (*June—July*).

4 Sea Clover (*T. squamosum*). A hairy annual, growing up to about 1 foot high. The flower-heads have short stalks, and usually a pair of leaves at the base with hairy stipules. The spiny calyx teeth spread outwards after flowering. It is a rare salt-marsh species found near the coast of south England and Wales. (*June—August*).

5 Strawberry Clover (*T. fragiferum*). This creeping Clover gets its name from the distinctive rounded fruiting heads; as the fruit ripens, the pinkish calyx of each flower becomes swollen. The hairless stems grow along the ground, rooting at the nodes, and spreading in meadows and grassy places, especially on heavy soils. It is locally common, though rare in the north. (*July—September*).

6 Alsike Clover (*T. hybridum*). A fairly large, hairless Clover, growing up to 2 feet tall. The individual flowers in the flower-heads are stalked, and the flower stalks bend downwards after flowering. This Clover was introduced into Britain as a crop plant, but is now common growing wild in fields and by roadsides throughout Britain. (*June—September*).

Clustered Clover (*T. glomeratum*). An uncommon annual, much smaller than Alsike Clover, with tiny, unstalked purple-pink flower-heads, growing at the axils of the leaf stalks. This is usually a prostrate plant of grassy places in the south and east, especially near the sea. (*May—June*).

Another rare Clover (*T. suffocatum*) grows in similar places. It is a tiny prostrate plant with triangular leaflets on upright stalks and unstalked pale pink or whitish flower-heads crowded together at the bottom of the leaf stalks.

1 **1 RED CLOVER**

2 **2 ZIGZAG CLOVER**

3 **3 HARE'S-FOOT CLOVER**

4 **4 SEA CLOVER**

5 **5 STRAWBERRY CLOVER**

6 **6 ALSIKE CLOVER**

LIFE SIZE

ROSACEAE: ROSE FAMILY (See p. 212)

1 **Dog Rose** (*Rosa canina* agg.). A very common shrub in hedges and thickets, with hooked prickles on the stems. There are 2 flat stipules, edged with tiny, pin-shaped glands, attached to the bottom of the leaf stalk. The toothed or lobed sepals usually bend backwards after flowering and fall off before the fruit ripens. The sweetly-scented pink or white flowers have many stamens, and are visited by insects for the pollen: there is no nectar. The styles are not joined together as in Field Rose (p. 78). The red hips are formed from the swollen receptacle (top of the flower stalk), and inside are the hairy fruits (achenes) (Fig. 1). During the Second World War, when oranges were scarce, many tons of hips were collected to make rose-hip syrup, rich in vitamin C. There are several different species of Dog Rose, and they are not easy to distinguish. (*June—July*).

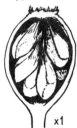

Fig. 1

Downy Rose (*R. villosa* agg.). This also common Rose is less tall than Dog Rose and has straight or only slightly curved prickles and deeper pink flowers. The leaves and hips are usually closely covered with short hairs, and the sepals do not usually turn backwards after flowering, as those of Dog Rose do, and may remain until the fruit is ripe. (*June—July*).

2 **Sweet-briar** (*R. rubiginosa* agg.). This small shrub can be easily recognized by the aromatic scent produced by rubbing the leaves. The prickles on the stems and flower stalks are mostly hooked. The flowers are usually bright pink, rarely white, and the orange-red hips are rounded. Sweet-briar is common in woods and hedges, and especially in scrub on chalky soil, but rare in Scotland. (*June—July*).

3 **Blackberry** or Bramble (*Rubus fruticosus*). There are several hundred species and hybrids in the *Rubus* group, and only an expert can identify all of them. A general description of the group is given here, and the white-flowered form is illustrated on p. 79. Cloudberry (p. 81), Dewberry, and Raspberry (p. 79) also are of the *Rubus* genus.

Blackberry has stems which either climb by prickles over hedges and bushes or trail along the ground, often sending out roots where the tips touch the ground. The leaves, some of which persist through the winter, are usually divided into 3 or 5 toothed leaflets, often with a whitish felt underneath. The flowers have numerous stamens and carpels. The fruit changes colour from green, through red, to black. Blackberry is common in woods, hedges, and scrub throughout Britain, especially in England and Wales. (*June—August*).

4 **Marsh Cinquefoil** (*Potentilla palustris*). This, the only purple-flowered British *Potentilla*, has a creeping rhizome and upright stems from 6 to 18 inches high. Both leaf stalks and flower stalks are hairy, and sometimes also the under side of the leaves. There are small green bracts below the flower, and 5 large, purple sepals alternate with the smaller petals. There are many stamens and carpels. This is a locally common plant of marshy and boggy places, especially in hills, though not common in the south. (*May—July*).

5 **Water Avens** (*Geum rivale*). A hairy perennial, up to 2 feet high, with the stem leaves much smaller and often less divided than the lower leaves. A few nodding flowers, which do not open wide, are borne at the top of the stem. The 5 calyx lobes alternate with 5 narrow bracts. The hooked styles remain attached to the hairy achenes. Except in the south, the plant is fairly common in damp, shady places. (*May—August*).

1 DOG ROSE   2 SWEET BRIAR   3 BLACKBERRY
4 MARSH CINQUEFOIL   5 WATER AVENS

LIFE SIZE

ERICACEAE: HEATHER FAMILY (See p. 214)

1 **Bell Heather** (*Erica cinerea*). A low evergreen shrub, up to 2 feet high, with branched stems rooting at the base. The smooth leaves are in whorls of 3, with many short shoots in the axils of the leaves, often giving the appearance of a bunch of leaves. The 8 stamens, which are hidden inside the bell-shaped corolla tube, have 2 toothed flaps below the anthers (Fig. 1*a*). The flowers are pollinated by bees. The fruit is a capsule surrounded by the corolla tube. This Heather is common on heaths and moors throughout the British Isles, often growing with Ling. (*June—September*).

Irish Heath (*E. mediterranea*) is a much larger shrub than Bell Heather, with smooth leaves in whorls of 4, and without short shoots in the leaf axils. The flowers in a one-sided spike have tubular corollas out of which project the purple anthers. It grows locally in bogs and heaths in Mayo and Galway in Ireland. (*March—May*).

2 **Dorset Heath** (*E. ciliaris*). An uncommon, much-branched, low shrub with hairy stems and hairs fringing the leaves, which are usually in whorls of 3. The stamens are hidden within the corolla tube but are without the toothed flaps which those of Bell Heather have (Fig. 1*b*). The style projects beyond the corolla tip. This Heath is found locally in Dorset, Devon, and Cornwall, and also in the west of France, Spain, and Portugal. (*June—September*).

Fig. 1

3 **Cross-leaved Heath** or Bog Heather (*E. tetralix*). The stems and upper leaf surfaces of this shrub are covered with short hairs. The leaves are in whorls of 4, making a cross when looked at from above. The flowers, in clusters at the ends of the stems only, have hairy sepals and 2 flaps below the anthers, as has Bell Heather (Fig. 1*a*). The flowers are pollinated by bees; some bees reach the nectar by making a hole near the bottom of the corolla tube, and so do not pollinate the flower. Cross-leaved Heath is common in bogs and moist places in heaths throughout the British Isles. (*June—September*).

Another species, Cornish Heath (*E. vagans*), has whorls of 4 or 5 leaves, and pale pink flowers spreading further down the stem and with protruding purple anthers. It is common near the Lizard in Cornwall, but not elsewhere.

4 **Bog Rosemary** or Marsh Andromeda (*Andromeda polifolia*). This small evergreen shrub up to 1 foot high has leaves which are shiny above and covered with a white bloom below. The calyx is 5-lobed, the corolla tube has 5 tiny teeth, and there are 10 stamens, each with 2 horns at the top of the anthers. The fruit is a capsule, opening in the middle by 5 slits. This is a rare plant of bogs, mostly in the north. (*May—September*).

5 **Ling** or Heather (*Calluna vulgaris*). A small shrub, usually less than 2 feet high, with branched stems rooting at the base. The name *Calluna*, from a Greek word meaning 'to brush', was given because the plants were once used to make brooms. The tiny, opposite leaves may be smooth or hairy. Ling, in contrast to other Heathers, has sepals that look like petals and are longer than the true petals. The flowers, which are sometimes white, have 8 stamens, each of which has 2 slender flaps at the base of the anthers as those of Bell Heather have (Fig. 1*a*). The flowers are visited by bees and other insects. Ling covers large stretches of heath and moorland throughout Britain, especially in the north and east and in east Ireland. In these places bees make the famous heather honey — in fact, bees are often taken there at the time when Ling is flowering. (*July—September*).

1 Bell Heather   2 Dorset Heath   3 Cross-leaved Heath
4 Bog Rosemary   5 Ling

LIFE SIZE

1 **Bearberry** (*Arctostaphylos uva-ursi*, family *Ericaceae*). This evergreen plant grows along the ground and has untoothed leaves which turn red in winter. The older parts of the stems have a blackish bark. The flowers have 5 calyx lobes and a corolla with 5 small teeth, inside which are 10 stamens. This plant (called 'bear-grape' in both Greek and Latin) is fairly common on moors in Scotland, N. England, and parts of Ireland. Its green leaves are used in medicines for kidney trouble. (*May—July*).

Black Bearberry (*Arctous alpina*). A rare Scottish moorland plant, which differs from Bearberry in having thin, narrow, toothed leaves that turn crimson and then drop off in the autumn. The flowers are white or pinkish, and the fruit is black when ripe. (*May—August*).

2 **Bilberry**, Blaeberry, or Whortleberry (*Vaccinium myrtillus*). A small, much-branched shrub, up to 1 or 2 feet high, with flattened or squarish, green twigs and toothed leaves, which fall off in autumn. The flowers are produced singly in the axils of the leaves. The calyx forms a small unlobed cup at the bottom of the rounded corolla tube. The blue-black berries have a bloom like that on a plum, and have a delicious flavour. Bilberry is common in woods and on moors, often covering large areas in N. England and Scotland, but rare in the south-east. It has various other country names, such as Huckleberry and Whinberry. (*April—June*).

The much rarer Bog Bilberry (*V. uliginosum*) has rounded stems and untoothed leaves with a bloom on the under surface. The small pink flowers are in groups of 1 to 4 in the leaf axils.

3 **Cowberry** or Red Whortleberry (*V. vitis-idaea*). A low shrub, with the stems growing along the ground. The evergreen leaves are very slightly toothed near the tip and have dots (glands) on the under surface. The flowers are in small groups at the ends of the branches. There are 4 sepals and a bell-shaped corolla with 4 lobes. The fruit, a rather acid red berry, is edible. This is a common plant of northern moors and woods, but rare in the south. (*June—September*).

4 **Crowberry** (*Empetrum nigrum*, family *Empetraceae*). A creeping, Heather-like shrub, with the leaves crowded together on the stems. The edges of the leaves are curled underneath so that they look very narrow. Male and female flowers are on separate plants, in the axils of the leaves. Three long stamens stick out from the male flowers, and the pollen is carried by the wind. Crowberry is a common moorland plant in most parts of the British Isles, except in the south. (*April—June*).

5 **Cranberry** (*Oxycoccus palustris*, family *Ericaceae*). An evergreen plant, with up to 4 flowers borne on fairly long, hairy stalks at the end of the wiry, creeping and rooting stems. There is a calyx of 4 broad sepals, and the corolla has 4 spreading or downwardly-bent petals, joined only at the base. The 8 stamens are clustered round the central style. The ripe fruit is used to make cranberry sauce, jam, and tarts. The plant is scattered throughout most of Britain, in bogs and wet heaths. (*June—August*).

Small Cranberry (*O. microcarpus*) has slightly smaller, triangular leaves and flowers in ones or twos on hairless stalks. It is a rare plant of bogs on Scottish mountains.

1 BEARBERRY   2 BILBERRY   3 COWBERRY
4 CROWBERRY   5 CRANBERRY

LIFE SIZE

1 **Purple Loosestrife** (*Lythrum salicaria*, family *Lythraceae*). A common upright perennial, with stiff, hairy stems up to 3 or 4 feet high. It is not related to the Yellow Loosestrifes (*see* p. 27). The flowers, which have hairy, strongly ribbed calyxes, are of three kinds, each growing on different plants. One type has a long style and stamens half of medium length and half short; the second type has a style of medium length, with long and short stamens; and the third type has a short style, with long and medium-length stamens. A visiting insect picks up pollen onto its body from long, medium, or short stamens, which is later dusted onto the stigma of another flower with a style of the same length. This arrangement ensures cross-pollination. Loosestrife grows in marshy places and by riversides, except in the north of Scotland. (*June—September*).

2 **Common Valerian** or All-heal (*Valeriana officinalis*, family *Valerianaceae*). A rather unpleasant-smelling common perennial, varying from 1 foot to as much as 5 feet tall, and found in damp and shady places throughout the British Isles. The lowest of the much divided leaves have long stalks. There are 3 stamens, a 5-lobed corolla, and a calyx, at first very tiny, which opens out to give a feathery plume on top of the fruit. As the English name All-heal and the Latin name *officinalis* suggest (*see* p. 206), it was valued for its healing powers, and was one of the medicinal herbs grown by monks. The drug extracted from Valerian roots is used as a sedative in the treatment of some nervous diseases. (*June—August*).

Marsh Valerian (*V. dioica*). This much smaller and less common Valerian grows not more than 1 foot tall and produces long creeping stems as well as upright flowering ones. It has undivided lower leaves, and separate male and female flowers on different plants. It is more common in marshy places in the north than in the south. (*May—June*).

3 **Red Valerian** (*Kentranthus ruber*). This Valerian grows from 2 to 3 feet tall and has all its leaves undivided. There is a long spur at the base of the long corolla tube, and each flower has only one stamen. The flowers are pollinated by butterflies. The plant, which may also have white flowers, is an escape from gardens, but has now become naturalized on dry banks and old walls, especially in Cornwall. (*May—September*).

4 **Butterbur** (*Petasites hybridus*, family *Compositae*). The very large leaves, which are greyish underneath with cottony hairs, usually appear just after the flowering stems. Male and female flower-heads are produced on separate plants. The male florets have a tubular, 5-toothed corolla, 5 stamens joined by their anthers, and a small pappus of hairs. The female florets (middle picture) have a very narrow corolla, a 2-lobed style, and a conspicuous pappus of long, white hairs. Butterbur is common by rivers and in damp places throughout the British Isles. (*March—May*).

White Butterbur (*P. albus*), a smaller plant with scented white flowers, is a rare northern species. Winter Heliotrope (*P. fragrans*), with scented white or lilac flowers, occurs fairly commonly as an escape from gardens. It is smaller and flowers earlier than Butterbur.

5 **Foxglove** (*Digitalis purpurea*, family *Scrophulariaceae*). A tall biennial, 2 to 5 feet in height, with large lower leaves up to 1 foot long, the undersides of which are greyish, with soft white hairs. There are a few long hairs inside the corolla tube, 2 long and 2 short stamens, and a slender style. White-flowered plants are occasionally found. Other names for this common plant of dry hillsides and open places in woods on acid soils are Dead Man's Bells (the plant is poisonous) and Fairy Thimbles. A drug, digitalin, is extracted from the leaves and seeds and used to treat heart diseases. (*June—September*).

QUARTER LIFE SIZE

1 Purple Loosestrife   2 Common Valerian   3 Red Valerian
4 Butterbur   5 Foxglove

1 **Common Centaury** (*Centaurium minus*, family *Gentianaceae*). A small annual, up to about a foot high, usually with a single stem coming from a rosette of leaves. These leaves have 3 distinct veins from the base and are slightly larger than the stem leaves. The number of flowers in the cluster varies, and sometimes they are almost unstalked. This common plant of dry, grassy places and woods gets its name from one of the Centaurs of Greek mythology, who is said to have used the plant as a medicine. (*July—September*).

Seaside Centaury (*C. littorale*) is a smaller rosette plant with much narrower leaves and a cluster of unstalked flowers. It grows locally by sea coasts in Wales and the north.

Slender Centaury (*C. pulchellum*). This species, found near the sea, especially in the south, has no rosette of leaves. It is not more than 6 inches tall, and its spreading branches bear fairly broad leaves and small, short-stalked flowers.

2 **Dumpy Centaury** (*C. capitatum*). This dwarf annual has a rosette of leaves at the base, with one or several stems carrying dense clusters of unstalked flowers. It grows in grassy places and on dunes, mainly in the south. (*June—September*).

3 **Toothwort** (*Lathraea squamaria*, family *Orobanchaceae*). Toothwort is a parasite, that is, it has no green leaves and gets its food materials from other plants. Roots from its fleshy creeping stem become attached to the roots of trees such as Hazel and Elm. It is usually rather shorter than the related Lesser Broomrape (p. 138) and has short-stalked flowers and a 4-lobed calyx with broad, hairy teeth. The 4 stamens are attached to the 2-lipped corolla. It is found occasionally in suitable woods or hedges throughout Britain. (*April—May*).

A much more spectacular Toothwort (*L. clandestina*), with bright purple flowers, occurs in a few places as a parasite on Poplar and Willow roots.

4 **Chaffweed** (*Centunculus minimus*, family *Primulaceae*). A rather uncommon, tiny, hairless annual with short-stalked pink or white flowers in the leaf axils. The parts of the flower are usually in fours, and the fruit is a round capsule opening by a lid, as in Scarlet Pimpernel (p. 104). It grows in damp, sandy or gravelly places throughout Britain. (*June—August*).

5 **Common Dodder** (*Cuscuta epithymum*, family *Convolvulaceae*). This, like Toothwort, is a parasite, but is attached by suckers to the stem, not the roots, of its host plant. The thin, weak, leafless stems twine anticlockwise round plants, mainly Gorse and Heather. There are usually 5 (rarely 4) calyx and corolla lobes, with 4 stamens sticking out from the corolla tube and 2 styles longer than the ovary. This plant is locally common in suitable places, rarer in the north. (*July—October*).

A rare parasite of Nettles, Great Dodder (*C. europaea*), has the stamens within the corolla tube and styles shorter than the ovary.

6 **Small Bindweed** (*Convolvulus arvensis*). The cone-shaped pink or white flowers of this plant are quite distinctive. They are in ones or twos in leaf axils on the weak, twining stems. Bindweed may kill other plants by twining tightly round them, and it is a difficult weed to get rid of in gardens because the rhizomes are very long and break easily. It occurs in fields and by roadsides. (*June—September*).

7 **Bog Pimpernel** (*Anagallis tenella*, family *Primulaceae*). A small, creeping and rooting perennial, with flowers on fairly long stalks in the axils of the hairless, opposite leaves. The 5 stamens with hairy stalks surround the central style. The fruit is a round capsule opening by a lid, as in Scarlet Pimpernel (p. 104). Bog Pimpernel grows in bogs and wet places, where it is fairly common, except in the south-east. (*June—August*).

8 **Sea Milkwort** (*Glaux maritima*). A spreading, rather fleshy perennial, one foot or less in height. The lower leaves are opposite, but the upper ones are sometimes alternately arranged. The pink flowers have no petals — what look like petals are really sepals. The 5 stamens alternate with the sepals. It is a fairly common salt-marsh plant. (*June—August*).

LIFE SIZE

1 COMMON CENTAURY	2 DUMPY CENTAURY	3 TOOTHWORT
4 CHAFFWEED	5 COMMON DODDER	6 SMALL BINDWEED
7 BOG PIMPERNEL	8 SEA MILKWORT	

125

1 **Bistort** or Snakeweed (*Polygonum bistorta*, family *Polygonaceae*). The name Bistort means 'twice twisted' and refers to the appearance of the thick rhizome. The slender, unbranched flowering stems grow to 1 or 2 feet high. The lowest leaves are long, with winged stalks, and the stalks of the stem leaves form a sheath round the stem. As with all members of this family, there are no petals, and the pink or white colouring of the flower spikes comes from the sepals. Bistort is an uncommon perennial of damp meadows and woods, seldom found in the south-east. (*June—August*).

Alpine Bistort (*P. viviparum*) is a smaller plant of mountain pastures and rocks in the north. The leaf stalks are not winged, and the leaves are narrow. There are often swollen buds called bulbils, in the flower spike, which drop off and give rise to new plants.

2 **Knotgrass** (*P. aviculare*). A slender, much-branched annual, with weak stems which either spread along the ground or grow up amongst other supporting plants. The white stipules which encircle the stem and the small clusters of flowers in the angle between leaf and stem are characteristic features. Knotgrass is very common in fields, waste places, and sea-shores, and is often a tiresome weed. (*June—September*).

3 **Spotted Persicaria** or Red-legs (*P. persicaria*). This annual can usually be easily recognized by the red stems and the dark blotch in the middle of the leaves. The sheaths round the stems are fringed with long hairs. The flowering spikes are often in pairs. Spotted Persicaria is common by roadsides, in ditches, and on waste and cultivated land throughout the British Isles. (*June—September*).

Pale Persicaria (*P. lapathifolium*). A branched, rather hairy annual, resembling Spotted Persicaria and often with blotches on the leaves, but with green stems, only short hairs fringing the sheaths round the stem, and usually white or greenish flowers. This Persicaria i nearly as common as Spotted Persicaria and grows in the same kinds of places.

A smaller and less common species, *P. nodosum* has pink flowers, spotted stems, and often yellow dots (glands) on the undersides of the leaves.

4 **Amphibious Bistort** (*P. amphibium*). There are two forms of this perennial, one growing in water, with smooth, floating leaves and stems, and one growing on land, with slightly hairy leaves and stems. The spikes of honey-scented flowers are often in pairs, and have 5 stamens, and 2 styles joined together near the bottom. It is common in streams, ponds, and wet places throughout the British Isles. (*July—September*).

5 **Thrift** or Sea Pink (*Armeria maritima*, family *Plumbaginaceae*). A tufted perennial with rosettes of grasslike leaves and hairy flowering stems without leaves. The solitary flower-heads are surrounded by papery bracts, the lowest of which form a sheath round the top of the flower stalk just below the flower-head. All the parts of the individual flowers are in fives. Thrift is common on cliffs and in muddy and sandy places along the coast, and also on inland mountains. (*March—October*).

6 **Cat's-foot** or Mountain Everlasting (*Antennaria dioica*, family *Compositae*). A creeping perennial, rooting at the nodes and producing upright flowering stems up to 9 inches high. The flowering stems and undersides of the leaves are covered with woolly white hairs. Male (6b) and female (6a) flowers are on different plants. The female florets have very slender corolla tubes surrounded by long hairs (the pappus). The male florets have wider corolla tubes surrounded by short hairs thickened at the ends. This plant grows on mountain pastures in the north. (*June—July*).

1 BISTORT   2 KNOTGRASS   3 SPOTTED PERSICARIA

4 AMPHIBIOUS BISTORT   5 THRIFT   6 CAT'S-FOOT

LIFE SIZE

GERANIACEAE: CRANESBILL FAMILY (See p. 211)

1 **Herb Robert** (*Geranium robertianum*). This annual is very common in woods, hedgebanks, and on stony places throughout the British Isles. Both hairy stems and leaves are reddish and have an unpleasant smell. Herb Robert differs from other Cranesbills in the shape of its leaves and the rounded ends of the petals. The flowers, which grow in pairs, hang downwards at night and in bad weather. Small-flowered Herb Roberts can be mistaken for a much less common species, Little Robin (*G. purpureum*), which has narrower petals and more narrowly cut leaves, and is found in dry places near south-western coasts. (*May—October*).

2 **Shining Cranesbill** (*G. lucidum*). As the name suggests, this annual has smooth, shining stems and leaves. The petals are rounded, but unlike Herb Robert, the leaves are also rounded and less deeply cut into 5 lobes. This plant is locally common on walls, rocks, and waste places, becoming rare towards the north. (*April—August*).

3 **Bloody Cranesbill** (*G. sanguineum*). A perennial with a stout creeping rhizome and branched hairy stems, which grow either upright or trailing along the ground. Both sides of the leaves carry white bristly hairs. This plant can be recognized by its large, solitary flowers, with petals slightly notched at the end. Plants with white or pink petals are occasionally found. The Bloody Cranesbill, though not very common, is most often found in dry rocky places, especially on limestone, or on sand-dunes. (*June—August*).

Mountain Cranesbill (*G. pyrenaicum*). This rather rare species has hairy leaves less deeply cut than those of the Bloody Cranesbill, and the purplish flowers, which grow in pairs, are smaller and have deeply notched petals. In spite of its name, this plant grows in meadows, waste places, and by roads, mainly in south and east England.

4 **Cut-leaved Cranesbill** (*G. dissectum*). This rather coarse-looking common annual has soft downward-pointing hairs on the branched stems, and stiff hairs on the leaves. The flower stalks are shorter than the surrounding leaves. After flowering, the hairy carpels split apart at the bottom and curl upwards, as in many species of *Geranium*. (*May—August*).

Long-stalked Cranesbill (*G. columbinum*). This species can be distinguished from Cut-leaved Cranesbill by its slightly larger, pinkish-purple flowers which grow on stalks that are longer than the surrounding leaves. The carpels are not hairy. It grows in dry pastures, scrub, and on hedgebanks, and is locally fairly common, though rarer in the north.

5 **Common Storksbill** (*Erodium cicutarium*). A tufted, often rather sticky annual, with flowers bearing 10 stamens, only 5 of which have anthers. As the fruit ripens, the long beaks of the carpels become twisted, so that the carpels, containing the seeds, break apart. The plant grows fairly commonly in dry, grassy or sandy places in most parts of Britain. (*May—September*).

There are two other species of Storksbills, both much less common. Sea Storksbill (*E. maritimum*) is a sticky little plant with oval, toothed leaves and small, often pale, reddish-purple flowers in pairs. Musk Storksbill (*E. moschatum*) is larger and rougher and smells of musk. It has 2 to 8 bluish-purple flowers. Both species are found near the sea.

6 **Dove's-foot Cranesbill** (*Geranium molle*). A very common annual, with branched, spreading stems and both stems and leaves covered with soft white hairs. The flowers vary in colour from reddish-purple to whitish. It grows on both waste places and cultivated land, but is rare in Scotland and Ireland. (*April—October*).

Small-flowered Cranesbill (*G. pusillum*). This is like a small Dove's-foot Cranesbill but with more deeply lobed leaves, flowers having only 5 of the 10 stamens with anthers, and hairy carpels. It is rather less common than Dove's-foot Cranesbill. (*May—October*).

Round-leaved Cranesbill (*G. rotundifolium*). This differs from the last two species in having less deeply lobed leaves and unnotched pink petals. It is rare, and grows on old walls and stony waste places, mainly in the south. (*June—August*).

LIFE SIZE

1 HERB ROBERT   2 SHINING CRANESBILL   3 BLOODY CRANESBILL
4 CUT-LEAVED CRANESBILL   5 COMMON STORKSBILL   6 DOVE'S-FOOT CRANESBILL

1 Comfrey (*Symphytum officinale*, family *Boraginaceae*). A hairy, rather rough perennial from 2 to 4 feet high, with thick roots and branching stems. The lowest leaves are long and stalked, the upper ones unstalked and running down the side of the stem. The flowers, which may be pink, mauve, cream, or white, have a 5-lobed corolla tube, inside which are 5 triangular scales alternating with the stamens. The fruit consists of 4 black, shiny nutlets. A herb tea for chest afflictions used to be made of the leaves, and the roots were used for poultices for sores and bruises. Comfrey grows in damp, shady places, especially by streams, throughout Britain, although less commonly in the north. (*May—July*).

A less common and larger species, Blue Comfrey (*S. peregrinum*), is found fairly frequently in woods and by roadsides. It has pink flowers that later turn blue. A more local species, mainly of the north, is Tuberous Comfrey (*S. tuberosum*), which is much smaller and less branched, and its leaves do not run down the stem. The flowers are yellowish-white.

2 Common Mallow (*Malva sylvestris*, family *Malvaceae*). A perennial, from 1 to 3 feet tall, with spreading hairs on the stems and leaves, and 3 small bracts below the 5-lobed calyx. The numerous stamens joined together in the centre of the flower bend downwards after they have shed their pollen, exposing the styles. The disk-shaped fruits are sometimes called 'cheeses' because of their shape (Fig. 1). This Mallow is common by roadsides and in waste places, especially in the south. (*June—September*).

x2

Fig. 1

Another Mallow, Dwarf Mallow (*M. neglecta*), is a smaller, more hairy plant, with prostrate growth and small, blue or lilac flowers.

A much taller and less common species, Marsh Mallow (*Althaea officinalis*), has very soft, greyish, only slightly lobed leaves, a ring of bracts below the calyx with 6 to 9 lobes, and large, soft pink flowers. It grows in dykes and drier salt-marshes near the sea. (*August—September*).

3 Musk Mallow (*M. moschata*). A species easily distinguished from Common Mallow by its deeply divided stem leaves. The petals are very broad at the ends and the fruit is hairy and rounded. Musk Mallow is fairly common on banks and by roadsides. (*June—September*).

Tree Mallow (*Lavatera arborea*). This stout biennial with stems woody at the base, may grow up to 10 feet tall. The soft, hairy leaves are slightly lobed, and the 3 long bracts below the calyx are joined together at the base. The petals are purplish, with deeper purple veins, and the yellowish fruits have wrinkled surfaces. This rather uncommon plant grows along sea coasts, mainly in the south-west. (*July—September*).

4 Deadly Nightshade (*Atropa belladonna*, family *Solanaceae*). A rather rare, very poisonous plant, often found near ruins. It was at one time cultivated by monks, who extracted from its roots and leaves a drug which acts on the heart, muscles, and nervous system. The rather surprising name *belladonna* ('fair lady') may have arisen because the plant is also used to make a cosmetic. Deadly Nightshade may grow to 3 feet or more in height and often bears pairs of leaves of unequal sizes. It is found locally in woods and waste places on chalky soil in England and Wales. (*June—August*).

5 Duke of Argyll's Tea-plant (*Lycium halimifolium*). This bushy, spiny shrub grows up to 6 or more feet tall, with stalked leaves about 2 inches long. The flowers are on short shoots with smaller, unstalked leaves. The fruit is a red berry. This plant, which is grown in gardens, has become naturalized on walls and waste places. (*June—September*).

6 Woody Nightshade or Bittersweet (*Solanum dulcamara*). A straggling perennial, woody at the base, with the lower leaves often 3-lobed. The flowers are like those of the Potato, to which it is closely related. Woody Nightshade is common in woods and hedges throughout Britain, except in the far north. It is poisonous, and a drug is extracted from the stems. (*June—September*).

Black Nightshade (*S. nigrum*), also poisonous, is a related species shown on page 95.

HALF LIFE SIZE

1 Comfrey	2 Common Mallow	3 Musk Mallow
4 Deadly Nightshade	5 Duke of Argyll's Tea plant	6 Woody Nightshade

PAPILIONACEAE: PEA FAMILY (See p. 211)

1 Everlasting Pea (*Lathyrus sylvestris*). A straggling perennial with a smooth, winged stem, climbing by means of its large branched tendrils to a length of 4 feet or more. Hidden within the 'keel' of the flower are 10 stamens, 9 of which are joined together by their stalks. The pods are 2 to 3 inches long, with numerous purplish-black seeds. This is a fairly common plant of hedges and woods. (*June—September*).

Earth-nut Pea (*L. tuberosus*). A rare plant of cornfields and waste places, especially near Fyfield, Essex: in fact, it is sometimes called the Fyfield Pea. The stem is not winged but angular, and the pairs of leaflets are more rounded than those of the Everlasting Pea. There are 2 to 5 flowers with crimson petals, and the pods are about 1 inch long. It has small tubers on the roots, from which it gets its name. (*June—August*).

Marsh Pea (*L. palustris*). This rare species differs from the previous two in that the leaves consist of 2 or 3 pairs of narrow leaflets, ending in a tendril (Fig. 1). The stems are winged, the flowers bluish-purple, and the pods about 1 inch long. It is found locally in fens and damp places, but not in Scotland. (*May—August*).

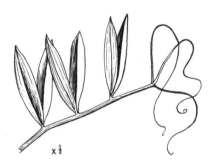

Fig. 1

2 Bitter Vetch or Tuberous Pea (*L. montanus*). A perennial with a creeping rhizome, swollen in places to form tubers. There are no tendrils, and the stem grows from 6 to 12 inches high. The flowers fade from bright reddish-purple to blue or greenish-brown. Each pod contains 4 to 6 brown seeds. This plant is most common in woods and hedges in hilly country, especially in the west and north. (*April—July*).

A very rare species, Black Pea (*L. niger*), is found in the mountains of Scotland. It has no tendrils, 6 to 10 leaflets, hairy flower stalks bearing reddish-purple flowers, and a hairy calyx.

3 Restharrow (*Ononis repens*). A hairy perennial with a woody, often creeping and rooting stem, which is sometimes spiny — 'arresting the harrow'. Stipules covered with sticky hairs are attached to the leaf stalks. The 10 stamens are joined together in one bundle (see No. 1). The calyx becomes enlarged in fruit and is longer than the pod, which contains 1 to 4 seeds (Fig. 2a). Restharrow is a common plant, especially on chalky grassland. (*June—September*).

Spiny Restharrow (*O. spinosa*) has more upright, sharply spiny stems, up to 18 inches tall, and 2 lines of hairs. The pod is usually longer than the calyx (Fig. 2b). It is fairly common in sandy and chalky pastures. Another species, Small Restharrow (*O. reclinata*), is a small, sticky annual with small flowers and pods pointing downwards. It is found occasionally on the west coast.

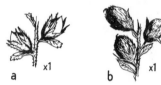

a   x1   b   x1

Fig. 2

4 Common Vetch (*Vicia sativa*). A straggling or climbing perennial with slightly hairy stems. The lobed stipules often have a dark spot in the middle. The pod is up to 2 inches long with 4 to 12 seeds. This Vetch is common in fields, by roadsides, and on waste ground through-out Britain. (*May—September*).

Two closely related species are the common Narrow-leaved Vetch (*V. angustifolia*), with smaller leaflets, flowers, and pods, and the rarer Spring Vetch (*V. lathyroides*), a small, often prostrate, somewhat hairy annual, with solitary, pale purple flowers.

Bithynian Vetch (*V. bithynica*). This can be recognized by its 1 or 2 pairs of large leaflets with long tendrils, and its solitary or paired bluish-purple flowers with white 'wings'. It grows on bushy cliffs near the sea, mainly in the south. (*May—July*).

5 Grass Vetchling (*Lathyrus nissolia*). This species has neither leaflets nor tendrils: the grass-like 'leaves' are really flattened leaf stalks. The solitary or paired flowers have very long stalks. The long pods, con-taining many roughened seeds, twist spirally when they ripen, and so split. This is a rather rare species of grassy places in the south and east. (*May—July*).

LIFE SIZE

1 Everlasting Pea   2 Bitter Vetch   3 Restharrow

4 Common Vetch   5 Grass Vetchling

PAPILIONACEAE: PEA FAMILY (See p. 211)

1 Tufted Vetch (*Vicia cracca*). A rather hairy perennial, climbing by its tendrils, often over hedges and bushes. It has small stipules at the base of the leaf stalk (Fig. 1a). The numerous flowers are borne on one side of the stem. The pods are long and flat, with several seeds. This Vetch is common in hedges and grassy places throughout the British Isles. (*June—August*).

Bitter Vetch (*V. orobus*). A similar species to Tufted Vetch but it does not climb, for instead of a tendril the leaf ends in a point (Fig. 1b). The flowers are paler and pinker, and the pods contain 3 or 4 seeds. It is a rare plant of woods and rocks, mainly in the north. (*May—July*).

Another rather rare Vetch is Wood Vetch (*V. sylvatica*), which differs from Tufted Vetch in having toothed stipules, oval leaflets (Fig. 1c), pale flowers, and broad pods. It is found locally in hilly woods and on cliffs by the sea.

2 Bush Vetch (*V. sepium*). A climbing perennial, with slightly zigzag stems, oval leaflets, and small, 2-lobed stipules. The flowers are few together, on very short stalks. The style has a tuft of hairs just below the stigma. The sharply pointed pods contain many seeds. This is a common plant of woods and hedges throughout Britain. (*May—September*).

3 Hairy Tare or Hairy Vetch (*V. hirsuta*). An annual, scrambling plant, which, in spite of its name, is often not hairy except for the small 2-seeded pods. The thin leaflets end in a small point in the centre of a notch, and the stipules are deeply cut. It is a common plant of fields and hedges in most parts of Britain, but rare in Ireland. (*May—August*).

Another similar plant, growing rather less commonly in the same places, is Smooth Tare or Slender Vetch (*V. tetrasperma*). It differs in usually having leaves with fewer leaflets, and hairless pods that, as the name *tetrasperma* suggests, are usually 4-seeded.

4 Lucerne or Alfalfa (*Medicago sativa*). This was introduced into Britain as a crop plant and has now become naturalized in some places. It is a perennial, 1 to 2 feet high, with slender, pointed stipules at the base of the hairy leaves. The very short-stalked blue or purple flowers are pollinated by bees. The spirally-twisted pods contain many seeds. (*June—August*).

Sainfoin (*Onobrychis viciifolia*). A perennial, up to 2 feet high, with Vetch-like leaves, except for an odd leaflet at the end of the leaf and no tendrils. The pink or red flowers are in groups on long stalks springing from the axils of the upper leaves (Fig. 2). The rough pod contains 1 seed. Sainfoin grows in grassy places on chalky soil, mainly in south-east England, and is also cultivated. (*May—September*).

Fig. 2   x½

5 Sea Pea (*Lathyrus maritimus*). A perennial with a creeping underground stem (rhizome), and stems above ground which are angular and spreading. The leaves usually end in 1 or 2 tendrils, and have large, sometimes toothed stipules. Sea Pea is a rather rare, local plant of sea shingle, mainly in south-east England. (*June—August*).

a   x½   b   x½   c   x½

Fig. 1

LIFE SIZE

1 TUFTED VETCH   2 BUSH VETCH   3 HAIRY TARE
4 LUCERNE   5 SEA PEA

135

1 **Common Fumitory** (*Fumaria officinalis*, family *Fumariaceae*). A spreading or climbing plant with flowers in groups of usually more than 20. There are several varieties which are now usually described as separate species. The plant was once believed to come out of the ground without seed, and so was named *fumus terrae* (Latin for 'earth smoke'). In fact, it has a one-seeded round nut. It is common on cultivated land and dry waste places, especially in the east. (*April—October*).

There is a very similar, locally common species, *F. parviflora*, which has white or pinkish flowers tipped with blackish-red; and in eastern counties another species, *F. micrantha*, is found, which has furrows in the narrow leaf segments.

Ramping Fumitory (*F. capreolata*). A larger and less common plant than Common Fumitory, that often climbs up to 3 feet high. It has broader leaf lobes and bigger flowers which are more closely crowded together. The petals are white or cream with purple tips. It is found most often in hedges and on cultivated land in the west. (*May—September*).

There is a shorter species (*F. purpurea*), which has purple, less densely crowded flowers; and a quite common spreading or climbing species, *F. boraei*, which has a flowering spike shorter than its stalk and dark-tipped pink flowers.

2 **Vervain** (*Verbena officinalis*, family *Verbenaceae*). A perennial with tough, hairy stems up to 1 or 2 feet high, and hairy leaves. This is the only British member of the *Verbenaceae*, which is a mainly tropical family and includes the huge Teak tree of tropical forests. Vervain was once much used in medicines and also used to be mixed in love potions. It is locally common in waste places in England and Wales. (*July—September*).

3 **Sea Aster** (*Aster tripolium*, family *Compositae*). This distinctive and attractive plant may grow up to 3 feet tall, and is common in salt-marshes and on cliffs and rocks round the coasts of the British Isles. In one variety there are no purple outer florets. (*July—October*).

4 **Lesser Periwinkle** (*Vinca minor*, family *Apocynaceae*). A perennial with creeping stems, rooting at the nodes. The petals are twisted round to the left in the bud. The 5 stamens, which alternate with the petals, have hairy tips, and there is a tuft of white hairs at the top of the style. Periwinkles, both the Greater and Lesser, occur locally in woods and copses. (*March—June*).

The Greater Periwinkle (*V. major*) is a larger plant with bigger flowers, and with the sepals edged by long hairs.

5 **Sea Lavender** (*Limonium vulgare*, family *Plumbaginaceae*). A perennial, with flowering stems branched above the middle, up to about 1 foot in height. This plant is related to Statice, everlasting flowers in which the coloured calyx persists after the petals have dropped off. These flowers can be dried and used for decoration in winter. Sea Lavender is not related to true Lavender, which is not a native of Britain. Sea Lavender is locally common in salt-marshes, often covering large areas. (*July—September*).

A similar but less common species, Lax-flowered Sea Lavender (*L. humile*), branches lower down the stem and has longer flowering spikes with the flowers further apart. Rock Sea Lavender (*L. binervosum*) also branches near the base, but can be distinguished by its smaller, 3-veined leaves. It is a rather rare plant of sea-cliffs and rocks.

1 Common Fumitory   2 Vervain   3 Sea Aster

4 Lesser Periwinkle   5 Sea Lavender

LIFE SIZE

1 Field Madder (*Sherardia arvensis*, family *Rubiaceae*). A small annual, with spreading, branched stems up to 12 inches long, and both stems and edges of the leaves rough with small bristles. Long, leaf-like bracts surround the clusters of flowers. The long, slender corolla tube has 4 spreading lobes, and there are 4 stamens and a 2-lobed style. The 4 bristly sepals enlarge as the fruit develops. This is a common plant of cultivated land and waste places throughout Britain. (*May—October*).

2 Pasque Flower (*Anemone pulsatilla*, family *Ranunculaceae*). A perennial with a rosette of much divided leaves, cut into narrow segments. Long, silky hairs cover the leaves, flower stalks, and the outside of the large, purple, petal-like perianth segments. There are numerous stamens, the outer ones being modified to form nectaries, which the bees visit. The fruits are achenes, with long, feathery plumes formed from the styles (Fig. 1). This Anemone is found only occasionally on chalk grassland, but where it does grow, it often grows in large numbers. (*April—May*).

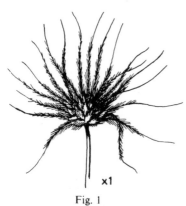

x1

Fig. 1

3 Lesser Broomrape (*Orobanche minor*, family *Orobanchaceae*). The Broomrapes are parasites, that is, they get their food material from other plants. Their tubers are attached to the roots of the host plant, and they produce flowering shoots without green leaves. They can be distinguished from the related Toothwort (p. 124) by their unstalked flowers and long, narrow calyx teeth. Lesser Broomrape is 6 to 18 inches high. The stigmas of the flowers are purplish-red. This Broomrape is usually a parasite on members of the Pea family, especially Clovers, or the Daisy family. It occurs mainly in the south. (*June—September*).

There are several other British Broomrapes, most of them rare. Greater Broomrape (*O. rapum-genistae*) grows on acid soils and may reach 3 feet high, with brownish flowers and yellow stigmas. It is a parasite

on Gorse and Broom, from which habit the group has got its name, meaning 'root-knob of broom'.

One species, Tall Broomrape (*O. elatior*), grows on Knapweeds on chalky soil, and has yellow flowers and stigmas. Another, smaller species, Red Broomrape (*O. alba*), has purplish-red flowers and grows on Wild Thyme, generally near the sea. Yet another, larger and very rare species (*O. caryophyllacea*), grows on Hedge Bedstraw in south-east Kent; it has pink or creamy flowers.

4 Common Butterwort (*Pinguicula vulgaris*, family *Lentibulariaceae*). This plant bears its flowers singly on stalks up to 6 inches high, which come from a rosette of fleshy, rather sticky leaves. Insects landing on the leaves may become stuck and are then digested by the plant. It grows in bogs and on wet rocks and is fairly common throughout the British Isles, except in the south. (*May—July*).

5 Pale Butterwort or Western Butterwort (*P. lusitanica*). A smaller plant than Common Butterwort, and the flowers are paler, with a shorter spur to the corolla. It is a rare and local plant of bogs in W. England, W. Scotland, and Ireland.

There is another rare species (*P. grandiflora*) which is found almost exclusively in S.W. Ireland, and which has magnificent large flowers, nearly an inch across.

6 Felwort or Autumn Gentian (*Gentianella amarella*, family *Gentianaceae*). A small, smooth biennial, between 3 and 12 inches high. The calyx and corolla have 5, rarely 4, equal lobes, and the corolla has a fringe of white hairs inside the top of the tube. The stamens, attached to the corolla, are arranged alternately with the corolla lobes. The fruit is a capsule. Felwort is fairly common on chalky pastures and old sand dunes. (*August—October*).

A tall, stouter, but rather rare species is *G. germanica*, with more rounded leaves, unequal calyx lobes, and showy lilac flowers. It is most often found on chalk grassland in the Chilterns.

7 Field Gentian (*G. campestris*). This Gentian differs from the Felworts in having the parts of its flowers in fours. The calyx is nearly completely divided into 2 narrow inner lobes and 2 broad outer ones. This Gentian is fairly common on dry grassland, except in the south-east. (*July—October*).

1 FIELD MADDER   2 PASQUE FLOWER   3 LESSER BROOMRAPE
4 COMMON BUTTERWORT   5 PALE BUTTERWORT   6 FELWORT
7 FIELD GENTIAN

LIFE SIZE

SCROPHULARIACEAE: SNAPDRAGON FAMILY (See p. 216)

1 Small Toadflax (*Chaenorrhinum minus*). A slender little annual, often downy and growing 3 to 6 inches tall. The flowers are borne singly on slender stalks in the angle between the stem and leaf (Fig. 1*a*). The fruit is a capsule, as are the fruits of most members of this family. It is fairly common in fields and waste places in the south, but rare in the north. (*June—September*).

Purple Toadflax (*Linaria purpurea*). This stiff, erect perennial is usually unbranched and grows up to 2 or 3 feet tall. It produces a spike of bright violet, Snapdragon-like flowers with curved spurs (Fig. 1*b*). It is found fairly frequently on waste ground and rough walls, mainly in the south. (*June—September*).

A less common species, which is usually found on chalky soils, is Pale or Creeping Toadflax (*L. repens*). This creeping perennial has branched flower stems carrying many purple-veined whitish flowers with an orange spot on the lower lip (Fig. 1*c*).

Weasel's Snout or Lesser Snapdragon (*Antirrhinum orontium*). This rather rare, small Snapdragon is an annual, up to 18 inches high, and has reddish-purple flowers with no spurs (Fig. 1*d*). It grows in cornfields and cultivated ground in the south, but not in the north. (*July—October*).

2 Figwort (*Scrophularia nodosa*). A stout, square-stemmed perennial, up to 3 feet high, with an unpleasant smell. The leaves are hairless, but small dots (glands) occur on the flower stalks. The flowers are pollinated by wasps. Figwort is common in damp, shady hedges and woods throughout the British Isles. (*June—September*).

Water Figwort or Water Betony (*S. aquatica*) is a taller, stouter plant, common in wet places and streamsides, mainly in the south. It has winged stems and leaf stalks, and leaves rounded at the tip.

3 Red Bartsia (*Odontites verna*). The stiff, upright stems grow to just over 1 foot in height, and stem, leaves, and calyx are covered with short, bristly hairs. The flowers have a tendency to face in the same direction. Red Bartsia is common in fields and by roadsides throughout the British Isles. (*June—September*).

Alpine Bartsia (*Bartsia alpina*) is a smaller, very rare plant found in mountain pastures and on rocks in the north. The upper lip of the purple corolla is much longer than the lower one.

4 Red-rattle or Marsh Lousewort (*Pedicularis palustris*). A much-branched annual, up to 2 feet high. The calyx is hairy, and there are 2 tiny teeth on each side of the hood-shaped upper lip of the corolla. The English name comes from the rattling sound of the ripe seeds inside the capsule. This plant is fairly common in wet places throughout Britain. (*May—September*).

5 Lousewort (*P. sylvatica*). This perennial, usually no more than 6 inches high, has spreading branches. The calyx is smooth outside, hairy inside, and there is only one tooth on each side of the corolla hood. It is a fairly common plant of marshes, bogs, and wet heaths. (*April—July*).

6 Ivy-leaved Toadflax (*Cymbalaria muralis*). This attractive little plant, common on walls throughout Britain, is easily recognized by its lobed leaves and Snapdragon-like flowers (Fig. 1*e*). It is pollinated by bees, and as the seeds ripen, the flower stalk bends towards cracks in the wall, into which the seeds fall when the capsule bursts open. (*May—September*).

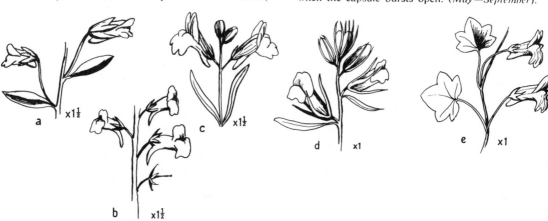

Fig. 1

1 SMALL TOADFLAX   2 FIGWORT   3 RED BARTSIA

4 RED-RATTLE   5 LOUSEWORT   6 IVY-LEAVED TOADFLAX

LIFE SIZE

LABIATAE: DEADNETTLE FAMILY (See p. 216)

1 **Corn Mint** or Field Mint (*Mentha arvensis*). A branched, rather hairy perennial, with a characteristic, pleasant, aromatic scent, and varying in height from a few inches to 2 feet. The flowers are in whorls, not at the ends of the stems, as with most Mints, but in the axils of the short leaf stalks. The hairy calyx has 5 broad teeth, and the stamens stick out of the 4-lobed corolla (Fig. 1*a*). This Mint is common in fields, waste places, and woods throughout the British Isles. (*July—October*).

Whorled Mint (*M. x verticillata*) is a commonly found cross between Corn Mint and Water Mint. It most resembles Corn Mint, but is often taller and stouter. The calyx teeth are narrow and pointed, and the stamens are within the corolla tube (Fig. 1*b*).

a   x2   b   x2

Fig. 1

2 **Basil-thyme** (*Acinos arvensis*). A small annual, usually about 6 inches high, with branched stems covered by short, soft hairs, and almost smooth leaves. The flowers, in groups of 3 or more in the axils of the leaves, have hairy calyxes, swollen at the base on one side. The 4 stamens are covered by the upper lip of the corolla. The flowers are visited by bees. It is a rather rare and local plant of fields and grassland, especially on chalky soil. (*June—September*).

3 **Horse-mint** (*Mentha longifolia*). A perennial with creeping underground stems and hairy upright stems from 1 to 3 feet high. The leaves are unstalked, whitish underneath, and hairy on both sides. The flowers, which are in long spikes at the end of the stems, are also hairy. Horse-mint grows in waste places and by streams and roadsides, but is only locally common. (*August—September*).

Apple-scented Mint or Round-leaved Mint (*M. rotundifolia*). This pleasantly-scented perennial, similar to Horse-mint but hairier, has broad, wrinkled leaves and pale pink flowers. It occurs locally by roadsides and waste places in England and Wales, but is rare elsewhere. (*August—September*).

4 **Water Mint** (*M. aquatica*). A very common, mint-scented, hairy perennial, easily recognized by the rounded heads of flowers at the ends of the stems and in the axils of the top one or two pairs of stalked and hairy leaves. The stamens stick out beyond the corolla tube. This plant grows in damp places of all kinds throughout Britain. (*July—October*).

Spear-mint (*M. spicata*). This is the Mint grown as a herb in gardens. It has unstalked, shiny leaves, and smooth, branched stems about 2 feet tall, with the flower spikes at the end. The stamens stick out from the corolla tubes. Spear-mint is fairly common by roadsides and in waste places, and spreads by creeping underground stems. (*August—September*).

Peppermint (*M. x piperita*). This is a hybrid between Water Mint and Spear-mint and is cultivated for its oils, which are used as a flavouring in sweets and medicines. The often reddish stems carry long, only slightly hairy leaves, with short stalks. There are long spikes of lilac flowers, and the stamens do not show beyond the corolla tube. Peppermint is found locally by streamsides and in damp waste places. (*August—September*).

5 **Penny-royal** (*M. pulegium*). A small perennial with often reddish, creeping stems, and flowering stems up to 1 foot high. Penny-royal has smaller leaves than the other Mints, and its calyx has 2 narrow teeth and 3 broader ones and a ring of white hairs just inside the calyx tube. It is a rather rare plant of damp heaths, mostly in the south. (*August—October*).

LIFE SIZE

1 CORN MINT   2 BASIL-THYME   3 HORSE-MINT
4 WATER MINT   5 PENNY-ROYAL

LABIATAE: DEADNETTLE FAMILY (See p. 216)

1 **Bugle** (*Ajuga reptans*). A low perennial with creeping and rooting stems and upright flowering stalks about 6 inches tall and hairy on 2 opposite sides. The bracts between the flowers are often coloured, and white- or pink-flowered specimens are occasionally found. Bugles differ from most members of the *Labiatae* family in having no upper lip to the corolla (Fig. 1). Bugle is common in damp woods and grassy places throughout Britain. (*May—July*).

There is another, rare Bugle, Pyramidal or Erect Bugle (*A. pyramidalis*), which has underground creeping stems and flowering stalks that are hairy on all sides. There are long, purple bracts between the pale blue flowers. It grows on limestone mountain rocks, mainly in Scotland.

Fig. 1

2 **Common Calamint** (*Calamintha ascendens*). This mint-scented plant grows up to 2 feet high. Both stems and leaves are hairy, and there is a ring of hairs inside the calyx. The 4 curved stamens are shorter than the upper lip of the corolla. This plant is, in fact, not very common, and is found on hedgebanks and waste places, mostly on chalky soil, but not in Scotland. (*July—September*).

A much rarer species, Lesser Calamint (*C. nepeta*), is found mainly in the south. It has small greyish-green leaves and paler blue or lilac flowers. A ring of hairs sticks up from the inside of the calyx tube.

3 **Lesser Skullcap** (*Scutellaria minor*). This is less hairy and smaller than Common Skullcap (see p. 177),

usually not more than 6 inches high. The bracts below the flowers are untoothed. It is locally common on wet heaths and boggy places, but absent from the north. (*July—October*).

4 **Self-heal** (*Prunella vulgaris*). A perennial with a creeping, rooting stem, and upright flowering stems from a few inches to about a foot high. Both stems and leaves carry tiny hairs, and there are longer ones on the calyx and flower bracts. The calyx is often purplish and the corolla occasionally white, though such flowers are not to be confused with Cut-leaved Self-heal shown on page 97. The plant was once thought to have medicinal properties, curing people without the help of a doctor — hence the name. It grows very commonly throughout the British Isles in woods and grassy places. (*June—August*).

5 **Wild Thyme** (*Thymus drucei*). A low, sweet-smelling perennial, with long, creeping stems. The squarish flower stems have dense hairs on 2 opposite faces (Fig. 2*a*), and the leaves are edged with hairs. The 4 stamens are as long as or slightly longer than the upper lip of the corolla. This plant is common on banks, heaths, and dry grassy places throughout the British Isles. (*June—September*).

A very rare related species of eastern England (*T. serpyllum*) has hairs on all sides of the flowering stems (Fig. 2*b*). The Larger Wild Thyme (*T. pulegioides*), a bigger plant found mainly on chalky soils in the south and east and usually not creeping, has hairs on the 4 corners of the flowering stems (Fig. 2*c*).

6 **Ground Ivy** (*Glechoma hederacea*). A creeping and usually hairy perennial with flowering stems up to 1 foot high. A forked style sticks out between the 2 lobes of the upper lip of the corolla and under this lip are the 4 stamens, the 2 outer ones shorter than the 2 inner ones. Ground Ivy is widespread and common on hedgebanks and waste ground and in woods, but becomes rarer in the north. (*March—June*).

a   b   c

Fig. 2

1 BUGLE   2 COMMON CALAMINT   3 LESSER SKULLCAP

4 SELF-HEAL   5 WILD THYME   6 GROUND IVY

LIFE SIZE

LABIATAE: DEADNETTLE FAMILY (See p. 216)

1 **Wild Basil** (*Clinopodium vulgare*). A softly hairy perennial, with square stems up to about 2 feet high, and short-stalked leaves, only slightly toothed. The whorls of flowers have very hairy, narrow bracts among them and 5-lobed hairy calyxes. The 4 stamens are shorter than the upper lip of the corolla, which does not form a hood. Wild Basil is common in hedges, by roadsides, and at the margins of woods in the south, less common in the north, and not found in Ireland. (*July—September*).

2 **Hedge Woundwort** (*Stachys sylvatica*). A tall, rather coarse perennial, with an unpleasant smell, and solid, square stems growing up to 3 feet high. Both stems and leaves have long white bristly hairs. The 4 stamens lie under the hooded upper lip of the corolla. The name Woundwort was given to several plants that were once considered to have healing properties and were used to make plasters and ointments. Hedge Woundwort is very common by roadsides, in woods, and on waste land throughout Britain. (*June—August*).

3 **Marsh Woundwort** (*S. palustris*). This differs from Hedge Woundwort in having hollow stems, narrower, usually unstalked leaves, short soft hairs, and paler flowers. It is widespread and common in ditches, fens, and on moist banks. (*July—September*).

Field Woundwort (*S. arvensis*). A small, hairy annual, with spreading branched stems less than 1 foot high. The rounded leaves with slightly toothed edges are about 1 inch long, and the upper ones are unstalked. The whorls of small, pale purple flowers form a loose spike (Fig. 1). This plant is a fairly common field weed in the west, though less common elsewhere. (*April—October*).

4 **Marjoram** (*Origanum vulgare*). A perennial, with reddish branched stems from 1 to 2 feet tall, and an aromatic scent. The small, almost untoothed, leaves are slightly hairy and dotted with glands on the under side. The bracts at the base of the flowers are often purple-coloured and longer than the 5-lobed calyxes. The 4 stamens stick out from the corolla. Marjoram is a common plant of hedgebanks and dry fields on chalky soil, though less common towards the north. It is also grown in herb gardens. (*July—September*).

5 **Betony** (*Stachys officinalis*). This species varies in height from about 6 inches to 2 feet. The stems carry only 2 or 3 pairs of leaves, of which the upper ones are almost unstalked. The lowest ones are larger, with long stalks. Most of the showy flowers are grouped in a head at the top of the stem, with a few in the leaf axils lower down. The calyx has 5 long, pointed teeth, and the 4 stamens are under the upper lip of the corolla. It is a common plant of hedgebanks, heaths, and woods, mainly in England and Wales. (*June—September*).

Fig. 1   x ½

2   3   4   5   *LIFE SIZE*

1 Wild Basil   2 Hedge Woundwort   3 Marsh Woundwort
4 Marjoram   5 Betony

LABIATAE: DEADNETTLE FAMILY (See p. 216)

1 Red Hemp-nettle or Narrow-leaved Hemp-nettle (*Galeopsis angustifolia*). An annual, varying in height from a few inches to over 2 feet, and recognizable by the narrow leaves, long bracts, and the long corolla tube, much longer than the calyx. Its Latin name, from *galea*, a helmet, refers to the shape of the corolla. This plant is fairly common in cultivated fields in most of Britain, but rare in Ireland. (*July—October*).

There is a rare, pale yellow-flowered relative, Downy Hemp-nettle (*G. dubia*), described on p. 28.

2 Common Hemp-nettle (*G. tetrahit*). A tall, hairy annual recognized by the stem swellings just below the nodes. White- and yellowish-flowered varieties are also found, and the corolla tube is only slightly longer than the calyx. The plant is common in fields and waste places throughout the British Isles. (*July—September*).

The less common Large Hemp-nettle (*G. speciosa*), described on p. 28, has yellow flowers with a purple lower lip.

3 Henbit Deadnettle (*Lamium amplexicaule*). The stems of this species grow along the ground and then bend upwards. The leaves (bracts) below the flowers are less rounded than the lower leaves, are unstalked, and clasp the stem less completely. This is a fairly common plant of cultivated land and waste places, though less so in Ireland. (*April—September*).

There is a northern species, *L. molucellifolium*, which has flowers more closely crowded together and very long calyx teeth.

4 Red Deadnettle (*L. purpureum*). This very common weed of gardens, cultivated land, and waste places throughout the British Isles has stems branched near the base, and grows from 6 inches to 1 foot high. The flower bracts are the same shape as the lower leaves and have short stalks. (*March—October*).

The locally common Cut-leaved Deadnettle (*L. hybridum*) is a more slender plant, with deeply cut leaves (Fig. 1) and a shorter corolla tube.

5 Black Horehound (*Ballota nigra*). A tall, hairy plant, up to 3 feet high, with an unpleasant smell. There are 2 or 3 sharp, narrow bracts at the base of each flower, and the calyx teeth are broad with pointed tips. Bees visit the flowers to collect the nectar produced at the bottom of the corolla tube. This plant is fairly common on hedgebanks in England and Wales, but rarer in Scotland and Ireland. It is one of many plants once recommended as a cure for the bite of a mad dog, and it was also used to relieve convulsions. (*June—October*).

Motherwort (*Leonurus cardiaca*). This rare perennial grows up to 4 feet tall and has a strong smell. It has whorls of pale pink or whitish flowers in 5-toothed calyxes up the stem at each leaf node. The leaves have 3 to 5 lobes at the end and taper towards the stalk (Fig. 2). Motherwort is usually found near buildings, and is used in nerve tonics and medicines for heart trouble. (*July—September*).

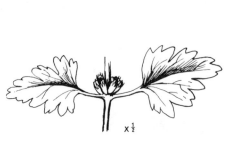

x ½

Fig. 1

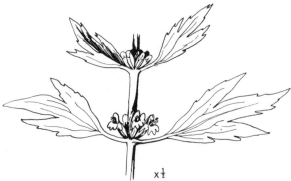

x ¼

Fig. 2

1 Red Hemp-nettle   2 Common Hemp-nettle   3 Henbit Deadnettle

4 Red Deadnettle   5 Black Horehound

LIFE SIZE

COMPOSITAE: DAISY FAMILY (See p. 218)

1 **Welted Thistle** (*Carduus crispus*). A biennial, up to 4 feet high, with very prickly leaves and leaf bases running down the stem, forming wings on either side of it. Both the stem and the under side of the leaves carry cottony hairs. The flower-heads, which are occasionally white and usually borne in small groups, are surrounded by narrow bracts. These and the pappus of long, unbranched hairs (Fig. 1*a*) on the fruits (achenes) distinguish this plant from Marsh Thistle (p. 153). The plant is common in fields, waysides, and waste places in the south; rarer in the north. (*May—September*).

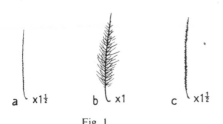

a `×1½`　　b `×1`　　c `×1½`

Fig. 1

2 **Woolly Thistle** (*Cirsium eriophorum*). A tall, stout biennial, 2 to 5 feet in height, with much-branched stems that are neither winged nor prickly. The stem leaves are unstalked, hairy underneath, and have spiny-tipped segments. The large flower-heads are usually carried singly, and are upright, with very conspicuous woolly hairs on the involucre. The pappus hairs are feathery (Fig. 1*b*). This rather rare Thistle grows locally in dry, grassy places on chalky soil, but is not found in Ireland or northern Scotland. (*July—September*).

3 **Musk Thistle** (*Carduus nutans*). This species has a winged stem up to about 3 feet high, sometimes branched near the top. The stem and spiny involucre are hairy, and there are a few hairs on the leaves. The flower-heads are usually solitary, long-stalked, and drooping, with a slightly musky smell. The pappus hairs are unbranched but rather rough (Fig. 1*c*). This is a fairly common Thistle of fields and waste places on chalky soils in the south, but rare in the north. (*May—September*).

4 **Spear Thistle** (*Cirsium vulgare*). A biennial, with stout stems up to 5 feet high, partly winged by the prickly

leaf bases. The leaves, which have bristly hairs on the upper side, end in long, sharp points. The flower-heads are carried singly, or 2 to 3 together, on short stalks. Unlike Musk Thistle, the spiny involucre is not woolly, and the pappus hairs are branched. Spear Thistle is very common in fields, waste places, and by roadsides throughout the British Isles. (*July—October*).

5 **Slender Thistle** or Seaside Thistle (*Carduus tenuiflorus*). An annual or biennial, varying in height from 1 to 4 feet. There are white hairs on the under side of the leaves and on the winged stems, which are branched near the top. The groups of small flower-heads, which are occasionally white, have rather long involucres of broad bracts with spiny tips. The pappus hairs are unbranched but rough. This is a fairly common Thistle of waste places, mainly near the sea. (*May—August*).

Milk Thistle (*Silybum marianum*). This rather rare annual or biennial Thistle has large, solitary, often drooping, purple flower-heads, surrounded by involucres of broad, leafy bracts with spiny tips (Fig. 2). The glossy green leaves have conspicuous white veins and spiny edges, and they clasp the stem but do not form wings down it. This Thistle is found locally in waste places, chiefly on limestone near the sea. (*June—August*).

×1

Fig. 2

LIFE SIZE

1 WELTED THISTLE   2 WOOLLY THISTLE   3 MUSK THISTLE

4 SPEAR THISTLE   5 SLENDER THISTLE

COMPOSITAE: DAISY FAMILY (See p. 218)

1 **Scotch Thistle** or Cotton Thistle (*Onopordum acanthium*). A very handsome, stout biennial, up to 5 feet tall, and covered with white woolly hairs. The long prickly stem leaves run parallel with and attached to the stem for much of their length, while the lower ones are lobed, with spiny teeth. The fruits are achenes, wrinkled and carrying a pappus of rough hairs. This Thistle grows in fields, waste places, and by roadsides but, in spite of its name, is rare in Scotland. (*July—September*).

2 **Creeping Thistle** or Common Field Thistle (*Cirsium arvense*). The creeping side roots give rise to upright stems, about 1 to 3 feet high, and carrying very spiny leaves. Male and female flowers are produced on separate plants, the female flower-heads (the back flower in the picture) having long bracts at their base and a feathery pappus of branched, brownish hairs. This is a very common Thistle throughout Britain and a difficult weed to get rid of because new shoots grow up from small pieces of root. (*July—September*).

3 **Marsh Thistle** (*C. palustre*). In contrast to Creeping Thistle this species has stems winged by the spiny leaves which have hairy upper surfaces, and there are more numerous flower-heads both at the ends of the stems and in the leaf axils. The male and female parts are in the same flower. It is common in marshes and damp woods and meadows throughout the British Isles. (*June—September*).

4 **Stemless Thistle** or Dwarf Thistle (*C. acaule*). A perennial with either a very short or no stem. There are long hairs on the veins on the under side of the leaves. The fruits are smooth achenes, carrying a pappus of long, whitish, branched hairs. This is a locally common species of chalky grassland in the south and may be a troublesome weed. (*July—September*).

5 **Meadow Thistle** or Marsh Plume Thistle (*C. dissectum*). A rather rare perennial with creeping underground stems and aerial stems from 6 inches to about 2 feet high. The leaves, which do not form wings down the stem, are toothed and greyish underneath with cottony hairs, and the lowest leaves are often deeply cut. This Thistle grows in damp meadows and fens in England and Wales as far north as Yorkshire. (*June—August*).

The Tuberous Thistle (*C. tuberosum*) has root tubers, leaves green on both sides, and the lowest ones divided into leaflets. It is a very rare and local plant growing on chalk grassland in the south.

Melancholy Thistle (*C. heterophyllum*). This tall, stout species resembles the Meadow Thistle but has large leaves which are white and felted underneath, toothed but not prickly, and with rounded lobes clasping the stem. The flower-heads are large and solitary, with reddish-purple or rarely white florets. It is a rather rare species of damp places on mountains in the north. (*July—August*).

LIFE SIZE

1 SCOTCH THISTLE   2 CREEPING THISTLE   3 MARSH THISTLE

4 STEMLESS THISTLE   **5 MEADOW THISTLE**

COMPOSITAE: DAISY FAMILY (See p. 218)

1 Lesser Knapweed or Hardheads (*Centaurea nigra*). A perennial, with tough, branching stems 1 to 2 feet high. The larger and lower leaves are slightly lobed or toothed. Knapweeds can be distinguished from the Thistles by their involucral bracts and their leaves, which are not prickly, and from Saw-wort by the comb of teeth round the top of the involucral bracts. The florets of Lesser Knapweed are either all alike, or the outer ones are larger and sterile. It is a common plant of waysides and meadows throughout the British Isles. (*June—September*).

Brown-rayed Knapweed (*C. jacea*) is a rare plant of S. England with paler, irregularly cut bracts round the large, purple flower-heads.

2 Saw-wort (*Serratula tinctoria*). A slender perennial, with stems 1 to 3 feet high, branching near the top and bearing a few flowers in clusters at the ends of the branches. The leaves, which may be divided into leaflets or almost undivided, have tiny teeth round the edge. The involucral bracts are untoothed and fringed by short hairs. The fruits (achenes) carry a roughish pappus of unbranched hairs. Saw-wort is a rather uncommon plant of meadows and open woods, especially on chalky soils. (*July—September*).

Alpine Saussurea or Alpine Saw-wort (*Saussurea alpina*). A rare mountain plant of Scotland, North Wales, and the Lake District. It has an unbranched flowering stem about 1 foot high, bearing a cluster of small flower-heads, with sweet-scented, purplish florets. The stem and the under side of the stalked, slightly toothed leaves are woolly. The fruit carries a pappus of short, unbranched outer hairs and long, branched inner ones. (*July—September*).

3 Greater Knapweed (*Centaurea scabiosa*). This is like Lesser Knapweed, but larger, with bigger, showy flower-heads which have spreading, sterile outer florets. The involucral bracts are green, with a comb of blackish teeth. It is common by roadsides and in meadows on chalky soil in the south, but rare in the north and in Ireland. (*July—September*).

4 Hemp Agrimony (*Eupatorium cannabinum*). This tall perennial looks rather more like a Valerian (p. 123) than one of the *Compositae*. The hairy stems may grow up to 4 feet high, with crowded clusters of reddish, mauve, or white flower-heads. Each individual head is small with about 5 florets, each having a 5-lobed corolla and a long, 2-lobed style (Fig. 1). This plant is common in damp woods and marshes and by rivers, except in the north. (*July—September*).

x2

Fig. 1

5 Blue Fleabane or Purple Fleabane (*Erigeron acris*). A small, rather hairy, upright annual or biennial, 1 foot or more high, branched near the top. The stalked leaves in a rosette at the base of the plant are rather more rounded than the stem leaves. The yellowish fruits (achenes) carry a pappus of long hairs spreading outwards like the spokes of a wheel. This Fleabane is locally common on walls and dry places in England and Wales. (*July—September*).

Canadian Fleabane (*E. canadensis*) is taller and less hairy than Blue Fleabane, with more numerous, smaller flower-heads. The outer florets are white or pinkish, the inner are yellow. This species grows on waste places and by roadsides, and as a weed in cultivated land, mostly in the south. (*August—September*).

6 Great Burdock (*Arctium lappa*). A tall, stout plant with very large leaves, up to 12 or 15 inches long, and covered with a grey felt underneath. The leaf stalks are solid and grooved. The hooks on the long-stalked flower-heads catch on the coats of animals, thus dispersing the seeds. The roots and leaves are used in medicines for skin diseases. Great Burdock is common by roadsides and in waste places, except in the far north. (*July—September*).

Lesser Burdock (*A. minus*) is a smaller but similar species, with hollow leaf stalks and almost unstalked flower-heads. Wood Burdock (*A. vulgare*) has hollow leaf stalks and large, stalked flower-heads.

1   2   3   4   5   6

LIFE SIZE

1 LESSER KNAPWEED	2 SAW-WORT	3 GREATER KNAPWEED
4 HEMP AGRIMONY	5 BLUE FLEABANE	6 GREAT BURDOCK

155

1 **Field Scabious** (*Knautia arvensis*, family *Dipsacaceae*). The members of this family, which in many ways resembles the *Compositae* or Daisy family, have flower-heads made up of a mass of tiny individual flowers each with a 4 or 5-lobed corolla, 4 long stamens, and cup-shaped calyx. The whole head is surrounded by rows of green bracts (involucre). The Field Scabious, also often called Bachelor's Buttons or Lady's Cushion, is common in dry grassland and cornfields and by roadsides, especially in the south. It is a hairy perennial, growing up to 3 feet or more in height, with a rosette of leaves, sometimes divided into leaflets and sometimes not. The tiny calyx cups surrounding the 4-lobed corollas have 8 slender bristles. The stamens are blue and burst to shed reddish pollen grains. (*June—September*).

2 **Small Scabious** (*Scabiosa columbaria*). This species is smaller and less hairy than Field Scabious, and has only one row of bracts surrounding the flower-heads. The calyx has 5 bristle-like teeth, and the corolla is 5-lobed. The large outer flowers are female only. The fruits, with the calyx still attached, are dispersed by the wind. This plant is fairly common on chalk downs and meadows in most parts of Britain. (*July—September*).

3 **Salad Burnet** (*Poterium sanguisorba*, family *Rosaceae*). A small perennial, from 6 inches to 1 foot tall, with a pleasant scent of cucumber when crushed. It is sometimes used to give a cucumber flavour to salads and sauces. Each flower-head contains 3 sorts of flowers: female at the top, bisexual (male and female) in the middle, and male at the bottom. There are no petals, greenish calyxes, numerous long-stalked stamens, and 2 styles with feathery stigmas. The flowers have no nectar, and pollen is carried by the wind. Salad Burnet is a common plant of waysides and grassland,

especially on chalky soil, but rare in Scotland. (*May—July*).

4 **Great Burnet** (*Sanguisorba officinalis*). This plant is taller than Salad Burnet, growing up to about 2 feet high. The flowers are all alike, with a crimson, 4-lobed calyx, 4 stamens, and a single style with a rounded stigma. This plant produces nectar, and so the flowers are pollinated by insects. Great Burnet grows on damp meadows in most parts of Britain, but is rather rare. (*June—August*).

5 **Teasel** (*Dipsacus fullonum*, family *Dipsacaceae*). A tall, handsome biennial with tough, prickly, hollow stems, sometimes as much as 6 feet high. The rosette of short-stalked oblong leaves at the bottom of the plant dies before flowering time. The pairs of stem leaves join round the stem at their base. The flowers, surrounded by spines, have tiny cup-shaped calyxes and 4-lobed corollas. The flower-heads often remain through the winter on the dead stems. This is a locally common plant of woods, hedges, and riverbanks in the south. (*July—September*).

A variety with spreading bracts below the flower-head and bracts with curved spines among the flowers used to be cultivated in parts of England for the woollen industry, the fruiting heads being used to comb the surface of the cloth and produce a nap. These are occasionally now found growing wild.

Small Teasel or Shepherd's Rod (*D. pilosus*). This is a smaller, less prickly plant, with a rosette of long-stalked leaves, and the stem leaves not joined together at the base. The corolla is whitish, and there are long hairs on the bracts between the flowers. It is found locally but infrequently in damp woods and on banks, especially on chalky soil, in Wales and southern England.

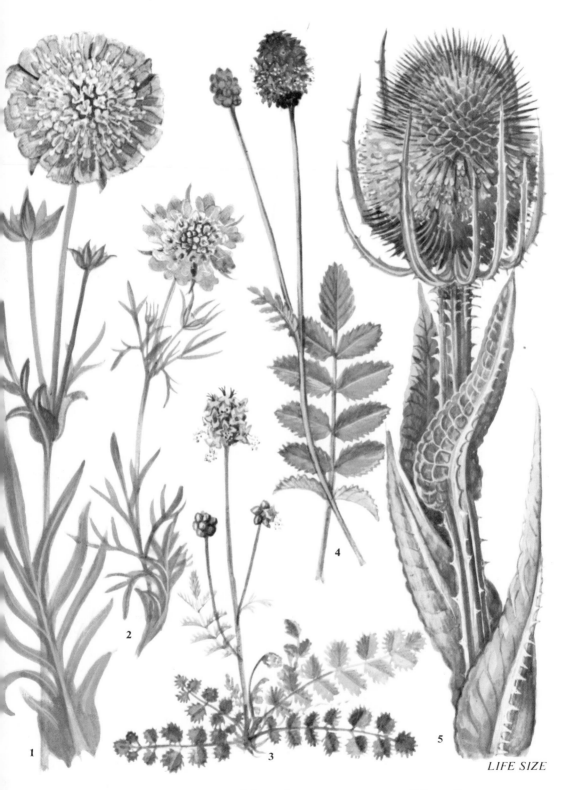

LIFE SIZE

1 Field Scabious   2 Small Scabious   3 Salad Burnet

4 Great Burnet   5 Teasel

ORCHIDACEAE: ORCHID FAMILY (See p. 219)

1 **Common Spotted Orchis** (*Orchis fuchsii*). A tall Orchis, growing up to 2 feet high, which can be recognized by the combination of solid stem, spotted leaves, and flowers crowded together in a pointed spike. The lower leaves are broad, becoming smaller and narrower up the stem until they look like bracts. The bracts below the flowers are narrow and fairly short. The flowers vary in colour from pale purple to white, with darker lines on the lower lip (Fig. 1*a*). This Orchis is found in wet grassy places, fens, and woods on chalky soil, mainly in the south and east and in Ireland. (*May—August*).

Heath Spotted Orchis (*O. ericetorum*) is a similar species to the Common Spotted Orchis, and is shown on page 101.

2 **Green-winged Orchis** (*O. morio*). This Orchis can be identified by its flower shape: all the sepals and petals, except the lip, arch upwards to form a hood (Fig. 1*b*). The wing-like sepals have green veins. The flower colour varies from deep purple to almost white, and the leaves are never spotted. This is a fairly common plant of pastures and woods in England and Wales, particularly on chalky soil, but is rarer in the north. (*May—June*).

Dark-winged Orchis or Burnt Orchis (*O. ustulata*). A small, rather rare plant, 4 to 9 inches high, with a short flower spike, whitish at the bottom but deep purple at the top. The flower has a hood, a very short spur, and a 4-lobed lip. This species grows on dry chalky grassland, mainly in south-east England. (*May—July*).

3 **Early Purple Orchis** (*O. mascula*). This is one of the earliest flowering Orchises and can be distinguished from the Spotted Orchises by the less crowded flowers with longer, notched middle lobes to the lips, and long, stout spurs, usually curving upwards (Fig. 1*c*). The flowers sometimes have an unpleasant smell of cats. This Orchis is common in woods and meadows throughout Britain. (*April—June*).

4 **Pyramidal Orchid** (*Anacamptis pyramidalis*). A plant with unspotted leaves shaped like the keel of a boat and becoming smaller up the stem. The stems are up to 18 inches high with characteristically pyramid-shaped flower spikes, which have a strong unpleasant smell. The flowers, with long, slender spurs (Fig. 1*d*), are pollinated by butterflies and moths. This is a fairly common Orchid of dry pastures and dunes on chalk. (*June—August*).

5 **Marsh Orchis** (*Orchis strictifolia*). A tall Orchis with a hollow stem, keel-shaped, unspotted leaves, and long wide bracts below the lowest flowers (Fig. 1*e*). The colour of the flowers varies a great deal, from purple to pink or almost white. It is a common plant of damp meadows and marshy places throughout Britain. (*May—July*).

There are two other somewhat similar Marsh Orchises. *Orchis praetermissa* is most common in the south and east and can be distinguished from *O. strictifolia* by the dark green leaves, which are not keeled, and the reddish-purple flowers with stout spurs. The other species, *O. purpurella*, found in marshes and fens in the north, is shorter, with smaller leaves that are sometimes spotted. The reddish-purple flowers are closely crowded into a short spike, and the stem is nearly solid.

6 **Fragrant Orchid** (*Gymnadenia conopsea*). The stems, 1 or 2 feet tall, carry oblong, keel-shaped lower leaves and narrow upper ones, all unspotted. The bracts are about the same length as the flowers, which are fragrantly scented and have long and slender spurs (Fig. 1*f*). This Orchid is fairly common throughout Britain on downs and heaths and in pastures, especially on chalky soil. (*June—August*).

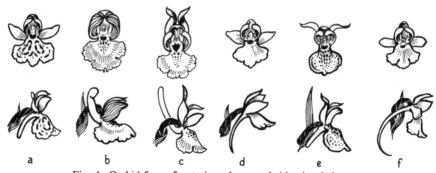

Fig. 1. Orchid faces, front view above and side view below
(*a*) Common Spotted Orchis. (*b*) Green-winged Orchis. (*c*) Early Purple Orchis.
(*d*) Pyramidal Orchid. (*e*) Marsh Orchis. (*f*) Fragrant Orchid.

LIFE SIZE

1 COMMON SPOTTED ORCHIS   2 GREEN-WINGED ORCHIS   3 EARLY PURPLE ORCHIS

4 PYRAMIDAL ORCHID   5 MARSH ORCHIS   6 FRAGRANT ORCHID

ORCHIDACEAE: ORCHID FAMILY (See p. 219)

1 **Broad Helleborine** (*Epipactis helleborine*). A solitary flower stem grows to 2 feet or more high from a short rhizome. The leaves are arranged in a spiral round the stem, the lowest ones being very broad. The flowers of all Helleborines, unlike most Orchids, have no spur, and the lip is divided into 2 parts by a fold in the middle. This species is found locally, mainly in beech-woods, throughout Britain. (*July—September*).

2 **Violet Helleborine** (*E. sessilifolia*). The clusters of usually purple stems grow up to 2 feet high. The rather narrow leaves are also often tinged with purple. The flower lip has a white end, curled under at the tip. The flowers are pollinated by wasps. This rather rare Helleborine grows in woods on chalky soil, mainly in beechwoods in the southern half of England. (*August—September*).

Narrow-lipped Helleborine (*E. leptochila*) is also a rare southern plant of beechwoods. The stems are sometimes several together, bearing numerous, fairly large, yellowish-green flowers, with longish, white-edged lips, coming to a point and not curled under. The leaves are in 2 rows, as in Dark-red Helleborine. (*June—August*).

Bog Orchid (*Hammarbya paludosa*). A small Orchid, only a few inches high, with a bulbous base, 2 to 4 broad leaves, and a spike of tiny, yellowish-green, stalked, spurless flowers, which are twisted so that the lip points upwards (Fig. 1). Though in general rare, Bog Orchid is locally common in bogs and wet moors, mainly in the north. (*July—September*).

×2

Fig. 1

3 **Fly Orchid** (*Ophrys insectifera*). The smooth stems may reach 2 feet tall, with a few large, oblong leaves. The 3 sepals are green, and the very long lip and 2 tiny, pointed petals resemble the body and first pair of legs of an insect. There is no spur. It is a rather rare plant of woods and grassy places on chalky soil, mostly in the south. (*May—July*).

Early Spider Orchid (*O. sphegodes*). A species somewhat resembling the Fly Orchid but with a broad, scarcely lobed flower lip, only slightly longer than the green sepals, and brown with paler markings (Fig. 2). This is a rare plant of chalky turfs in the south and east. (*April—May*).

×1

Fig. 2

4 **Marsh Helleborine** (*Epipactis palustris*). The solitary stem of this species grows from 6 to 18 inches high from a creeping rhizome. The narrow leaves, the lower more rounded than the upper, have pointed tips. The greenish sepals are hairy on the outside, and the lower part of the flower lip has frilly edges. The flowers are pollinated by bees. Though in general rare, this species is locally common in fens, though not in northern Scotland. (*June—August*).

5 **Bee Orchid** (*Ophrys apifera*). An Orchid easily recognized by the broad, velvety lip, which looks like a bee settling on the flower. From 2 to about 5 flowers are borne on the smooth stem, which is about 6 to 18 inches tall. The leaves are rather small, the upper ones resembling the bracts below the flowers. This rather rare plant of chalky soils grows in grassy places and at the edges of woods, more often in the south than the north. (*June—July*).

6 **Dark-red Helleborine** (*Epipactis atrorubens*). This plant has a solitary, hairy stem about 12 inches high, with folded leaves in 2 opposite rows. The unspurred, slightly scented flowers, with hairy ovaries, are pollinated by bees and wasps, and are found only locally in woods and rocky places on chalky soils, mainly in the north and west. (*June—August*).

1 BROAD HELLEBORINE   2 VIOLET HELLEBORINE   3 FLY ORCHID
4 MARSH HELLEBORINE   5 BEE ORCHID   6 DARK-RED HELLEBORINE

LIFE SIZE

1 **Gladdon** or Roast-beef Plant (*Iris foetidissima*, family *Iridaceae*). A stiff, upright perennial, about 1 to 2 feet tall, with an unpleasant scent. The narrow evergreen leaves are as long as or sometimes longer than the flowering stem, which carries 2 or 3 flowers. There are 6 perianth segments, the 3 outer ones being longer than the inner and curved downwards. The 3 yellowish, petal-like styles cover the 3 stamens. The capsules of orange-red seeds split open when ripe and are conspicuous during the autumn in woods, hedgebanks, and on cliffs. The plant is locally common in the south, especially on chalky soil. (*May—July*).

2 **Meadow Saffron** (*Colchicum autumnale*, family *Liliaceae*). This perennial plant flowers in the autumn, after all the leaves have withered — hence its country name, Naked Ladies. The long, slender perianth tube rises straight from the corm, which is a small, bulb-shaped underground stem. The flowers have 6 stamens with orange anthers, and 3 thread-like styles. The fruits appear with the narrow, glossy green leaves in the following spring. The plant is very poisonous to animals and may even kill them. It grows in occasional woods and damp meadows in England, Wales, and south-east Ireland, especially on limestone. The seeds and corms are collected for the extraction of a narcotic drug, colchicine, used especially to cure gout. (*August—October*).

3 **Autumn Crocus** (*Crocus nudiflorus*, family *Iridaceae*). Like Meadow Saffron, this species also flowers in the autumn after the leaves have died down. The short-stalked flowers have 3 conspicuous styles, which divide into lobes at the top. There are 3 pale yellow stamens. The tufts of narrow leaves, which appear in the spring, are green with a white stripe on the under side. This rare meadow plant grows mostly in the midlands and the north-west. (*August—October*).

The Purple or Spring Crocus (*C. purpureus*), which has become naturalized in meadows in parts of England, has purple or white flowers appearing with the leaves in the spring. (*March—April*).

4 **Fritillary** or Snake's Head (*Fritillaria meleagris*, family *Liliaceae*). This beautiful and increasingly rare plant is found chiefly in damp meadows in the Thames Valley and in the south-east. It is also grown in gardens and may be found as a garden escape elsewhere. The stems are from 9 to 18 inches high, with the nodding flowers borne singly or occasionally in pairs. (*April—June*).

5 **Crow Garlic** (*Allium vineale*). A bulbous perennial, with hollow, rounded leaves growing up to 2 feet high. The flowers, with long protruding stamens, are mixed with bulbils, swollen buds that fall off the parent and give rise to new plants. Sometimes there are no flowers, only bulbils. The plant has a smell of onions. It grows in fields and waste places and is fairly common in England and Wales, local in Scotland and Ireland. (*June—August*).

Field Garlic (*A. oleraceum*). The leaves of this rarer species are flattened and partly solid. The bracts below the flower-heads are much longer than the flowers (Fig. 1a). The pinkish or brownish flowers have no protruding stamens and are mixed with bulbils. This is a local plant of dry fields and waste places, mainly in the east. (*July—August*).

Chives (*A. schoenoprasum*), which is often cultivated in gardens, is occasionally found wild in dry, rocky places. The tufts of leaves are rounded, and the flowers are pinkish-purple, without bulbils (Fig. 1b). (*June—July*).

There are several species of Wild Leek, all of them very local and most of them producing bulbils. For the most part they are taller than the Garlics.

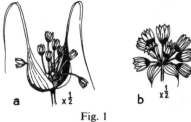

Fig. 1

6 **Turk's Cap Lily** (*Lilium martagon*). An erect, rather rough plant, 2 to 3 feet high. The flowering stem bears from 3 to 10 flowers, which are pollinated by butterflies. This handsome plant is a popular garden Lily, and occurs wild in woods, chiefly in the south. It also grows in the mountains of Central Europe and in parts of Russia and Greece. (*August—September*).

1   2   3   4   5   6

*THREE-QUARTERS
LIFE SIZE*

1 GLADDON   2 MEADOW SAFFRON   3 AUTUMN CROCUS
4 FRITILLARY   5 CROW GARLIC   6 TURK'S CAP LILY

VIOLACEAE: VIOLET FAMILY (See p. 210)

1 **Common Dog Violet** (*Viola riviniana*). A small, smooth or slightly hairy perennial, with unscented flowers. The flower stalks spring from the axils of leaves on side branches; the short main stem ends in a tuft of leaves. The small green stipules at the bottom of each leaf stalk are edged with fine teeth. The sepals are narrow, and the lowest petal forms a spur which stores nectar. Common Dog Violet can be distinguished from Wood Dog Violet (No. 7) by the spur, which is paler than the petals, curved, and notched at the tip. The fruit is a capsule that splits into 3 parts to release the seeds (Fig. 1). This Violet is common throughout the British Isles. (*March—June*).

×2

Fig. 1

2 **Wild Pansy** or Heartsease (*V. tricolor*). At the bottom of each leaf stalk are leafy, deeply cut stipules. The large, variously coloured flowers are sometimes purple or yellow only (*see* p. 7). This Pansy is fairly common in cornfields, grassy places, and waste ground throughout Britain. (*May—September*).

Field Pansy (*V. arvensis*), another common Pansy, especially in arable ground, has smaller flowers, with the yellow, white, or purplish petals shorter than the sepals.

Mountain Pansy (*V. lutea*). This species has larger flowers than has Wild Pansy, with long spurs. They may be yellow, purplish-blue, or the upper petals purple and the rest yellow. One or two stems arise from a creeping rhizome. The small leaves have 3 or 4 teeth along each side and leafy, lobed stipules at the base. It is locally common in mountain pastures in the north. (*June—August*).

3 **Sweet Violet** (*V. odorata*). Unlike other British Violets the violet or white flowers of this perennial are scented. There are no branched stems, the flowers and slightly hairy leaves all coming from the base. Sweet Violet is common at the edges of woods and on hedgebanks, especially on chalky soils, and spreads by rooting runners. (*February—April*).

4 **Heath Dog Violet** (*V. canina*). This is like Common Dog Violet, but the flowers are bluer (occasionally white), the longer leaves are slightly hairy on the upper side, and there is no central tuft of leaves. The stipules at the bottom of the leaf stalks are very small. The sepals have pointed tips. This is a locally common plant of heaths, sand-dunes, and fens throughout Britain. (*April—June*).

The less common Pale Heath Violet or Milky Violet (*V. lactea*) has fairly large, toothed stipules, and very pale violet flowers with a yellowish or greenish spur. It flowers later than Heath Dog Violet.

5 **Hairy Violet** (*V. hirta*). This is like Sweet Violet, but has unscented, usually paler flowers and very hairy leaves. There are no runners. The ends of the sepals are rounded, instead of pointed as in the Dog Violets. Hairy Violet is common on chalk grassland and in woods, but rare in Ireland. (*March—May*).

6 **Marsh Violet** (*V. palustris*). A perennial with a creeping stem and hairless leaves that are more rounded than those of other Violets. The flower has a short, broad spur and sepals with rounded ends. It grows in fens, marshes, and wet woods and is more common in the north than in the south. (*April—July*).

Fen Violet (*V. stagnina*). A rare plant of fens, mainly in the east. The leaves, longer and narrower than those of the Marsh Violet, are borne on shoots which grow up from a creeping underground stem. The pale purplish or white flowers have short greenish spurs and pointed sepals. (*April—June*).

7 **Wood Dog Violet** or Pale Wood Violet (*V. reichenbachiana*). A small, hairless perennial with runners, which is much like Common Dog Violet except that the flowers have straight, slender, and unnotched spurs, darker than the petals. The flowers are lilac to reddish in colour, rarely white. This common plant of woods and hedges in the south is rarer in the north. (*March—May*).

1 COMMON DOG VIOLET   2 WILD PANSY   3 SWEET VIOLET
4 HEATH DOG VIOLET   5 HAIRY VIOLET   6 MARSH VIOLET
7 WOOD DOG VIOLET

LIFE SIZE

CAMPANULACEAE: BELLFLOWER FAMILY (See p. 217)

1 **Spreading Bellflower** (*Campanula patula*). A slender but rather rough plant, up to 2 feet high, with the leaves completely or almost unstalked. The branched flowering stem carries a few flowers on each branch, and these have 5 stamens and a hairy style with 3 stigmas. The fruit, a capsule, opens by tiny holes near the top. This is a rare plant of hedges, woods, and shady places, mainly in the west. (*July—September*).

A similar species, *C. rapunculus*, has unbranched or only slightly branched flowering stems, and smaller, pale blue flowers. The fleshy root was at one time eaten in salads. Though very rare, it is scattered over England and south Scotland in fields and hedgerows.

2 **Nettle-leaved Bellflower** or Bats-in-the-belfry (*C. trachelium*). A perennial, from 1 to 3 feet high, with bristly hairs on the leaves, on the squarish stems, and on the deeply toothed calyxes. The lowest leaves are long-stalked, but the stem leaves have only very short stalks. The flowers are borne singly or in groups of 2 or 3. This is a fairly common plant of woods and hedges on heavy soils throughout Britain. (*July—September*).

3 **Clustered Bellflower** (*C. glomerata*). This Bellflower is so called because of the cluster of unstalked flowers at the end of the stem. The leaves at the base of the plant have long stalks while the upper ones clasp the stem; all are covered with soft hairs. It is fairly common in dry, chalky fields. (*May—September*).

4 **Creeping Bellflower** (*C. rapunculoides*). Upright stems grow about 1 or 2 feet high from creeping, underground stems. The basal leaves are broad and long-stalked; the upper ones are almost unstalked, and there are small, narrow bracts below the upper flowers. The hairy calyx teeth bend back when the blue or purple flowers mature. Creeping Bellflower is a rare plant of woods, railway embankments, and waysides, especially near houses. (*July—September*).

5 **Giant Bellflower** (*C. latifolia*). A large, slightly hairy perennial with grooved stems up to 4 feet high and long-stalked basal leaves. The stem leaves are unstalked. The blue or white flowers have a smooth calyx with long, narrow teeth pointing upwards. It is fairly common in woods in the north, rare in the south. (*July—August*).

6 **Harebell** or Bluebell (*C. rotundifolia*). A smooth, slender perennial with creeping underground stems. The lowest leaves are rounded (hence the Latin name), but these usually wither before flowering time. The calyx teeth are smooth and very narrow. Harebell, called Bluebell in Scotland, is common in heaths and dry places throughout the British Isles. (*July—September*).

Ivy-leaved Bellflower (*Wahlenbergia hederacea*). A slender, creeping stem carries small, long-stalked leaves, somewhat like young Ivy leaves (Fig. 1). The pale blue, bell-shaped flowers are borne singly on very long stalks in the leaf axils. This small and very local plant is found on damp heaths and in woods in the south and west. (*July—September*).

Fig. 1

Venus's Looking-glass (*Specularia hybrida*). A small, hairy annual bearing unstalked, oblong leaves with wavy edges, and blue or purplish flowers with outspread corolla lobes and long calyx teeth (Fig. 2). The fruit is a long capsule. This is a locally common plant of cornfields, mostly in the south. (*May—September*).

Fig. 2

1 SPREADING BELLFLOWER   2 NETTLE-LEAVED BELLFLOWER   3 CLUSTERED BELLFLOWER
4 CREEPING BELLFLOWER   5 GIANT BELLFLOWER   6 HAREBELL

LIFE SIZE

1 Pale Flax (*Linum bienne*, family *Linaceae*). A slender, hairless plant, growing to a foot or more tall and usually branching near the base. The parts of the flowers are in fives, with long, pointed sepals. The stamens are joined round the bottom of the ovary. The fruit, a capsule, has 10 teeth at the top when it has split open. This Flax is a rather rare and local plant of grassland on chalky soils in the south, especially near the sea. (*May—October*).

Common Flax (*L. usitatissimum*). This is the Flax grown for its oil-seed (linseed) and for its fibre for making linen. It is sometimes found as an escape from cultivation. It is unbranched and has larger flowers than Pale Flax. Another rare Flax, *L. anglicum*, is a tufted plant of chalk grassland in the east, with rounded sepals and large, bright blue petals. (*June—August*).

2 Meadow Cranesbill (*Geranium pratense*, family *Geraniaceae*). This has much bluer flowers than the other British Cranesbills (*see* p. 128). It is a rather hairy perennial, growing to 1 or 2 feet high. The lowest, deeply-lobed leaves have long, often reddish stalks while the upper stem leaves are smaller and almost unstalked. The red hairs on the long, pointed sepals and upper part of the stem have glands forming rounded tips. Meadow Cranesbill is fairly common in meadows and by waysides, but rare in Ireland and northern Scotland. (*June—September*).

Wood Cranesbill (*G. sylvaticum*). This is like Meadow Cranesbill but with less deeply-lobed leaves and usually more numerous, purplish-blue flowers. It grows in damp woods and fields, mainly in the north. (*June—August*).

Dusky Cranesbill (*G. phaeum*). A rather uncommon plant of hilly woods and hedgebanks, with a number of rather small, dark purple flowers. The lowest leaves are divided into broad, slightly-cut lobes, with long hairs on the upper surface. (*May—July*).

3 Bluebell or Wild Hyacinth (*Endymion nonscriptus*, family *Liliaceae*). The stems, containing a sticky white juice, grow up to 12 or 18 inches high, with the flowers borne on short stalks on one side. The 'bell' is formed by 6 perianth segments, and there are 3 long and 3 short stamens. Seeds are produced, but the plant also survives the winter as a bulb. It is common

and widespread, often making a blue carpet in woods. (*April—June*).

4 Spring Squill (*Scilla verna*). A small plant, up to 6 inches high, with narrow leaves, and flowers crowded together at the end of the leafless stem. There is a long, pale bract at the bottom of each short flower stalk. It is a rare plant of grassy places near the sea. (*April—May*).

Autumn Squill (*S. autumnalis*), an even rarer, later-flowering plant of southern coasts, has no bracts. (*July—September*).

5 Chalk Milkwort (*Polygala calcarea*, family *Polygalaceae*). The upper leaves are arranged alternately on the stem and are smaller than the rosette of lower short-stalked leaves near the bottom of the stem. There are 3 tiny and 2 large, petal-like sepals, and the blue, pink, or whitish petals are joined to the stamens. The fruit is a capsule, very broad when ripe (Fig. 1*a*). This rare species is found on chalk grassland in the south of England. (*May—July*).

6 Heath Milkwort (*P. serpyllifolia*). This slender plant differs from other Milkworts in having the lower leaves arranged opposite one another. The flowers are usually blue. It is common on grassland and heaths throughout the British Isles. (*May—August*).

7 Common Milkwort (*P. vulgaris*). A slender perennial, from a few inches to a foot high, with alternate leaves, far more pointed than those of Chalk Milkwort. The flowers are blue, pink (p. 113), or white, and the ripe capsule is not so broad as that of Heath Milkwort (Fig. 1*b*). This is a common grassland plant, especially on chalky soils. (*May—September*).

Another species (*P. oxyptera*) differs from Common Milkwort in being smaller, often growing along the ground on heaths and dunes, and in having broader fruiting capsules.

a   x2   b   x2

Fig. 1

LIFE SIZE

1 PALE FLAX   2 MEADOW CRANESBILL   3 BLUEBELL
4 SPRING SQUILL   5 CHALK MILKWORT   6 HEATH MILKWORT
7 COMMON MILKWORT

BORAGINACEAE: FORGET-ME-NOT FAMILY (See p. 215)

1 **Sea Lungwort** or Oyster Plant (*Mertensia maritima*). The stem and leaves of this perennial are covered with bloom like that on a plum. The plant grows along the ground. The leaves are thick, with dots on the upper surface. The 5 stamens stick out of the open corolla tube. This is a rare plant of stony sea-shores in the north. (*June—August*).

2 **Alkanet** (*Pentaglottis sempervirens*). A perennial up to 3 feet tall and rough with bristly hairs. The leaves, as the Latin name suggests, remain during the winter. The small clusters of flowers on short side branches have short corolla tubes, which are closed at the mouth by white scales forming a central 'eye'. The stamens are hidden inside. It is a rather local plant of waysides and near gardens, mostly in the south-west. Several different plants from the roots of which a red dye used to be extracted are called Alkanet. (*May—July*).

3 **Blue Gromwell** (*Lithospermum purpurocaeruleum*). A hairy, but not rough, perennial with long, creeping stems and flowering shoots 1 to 2 feet high. The 5 narrow sepals are joined only at the base, and the 5 stamens are hidden in the long corolla tube. Blue Gromwell is a rare woodland plant of chalky soils in the south and west. (*May—July*).

Narrow-leaved Lungwort or Joseph and Mary (*Pulmonaria longifolia*). A very rare plant of woods on clay soils in the south. It is rather similar to Blue Gromwell, but has very long, narrow, white-spotted basal leaves and smaller flowers with the calyx divided into 5 teeth at the top only. (*March—June*).

Common Lungwort (*P. officinalis*) is a rare garden escape in woods and hedges in the south. It has rounded leaves and slightly larger flowers than those of Narrow-leaved Lungwort.

4 **Viper's Bugloss** (*Echium vulgare*). A tall biennial, about 1 to 3 feet in height, covered with white bristly hairs. The rosette of large, stalked leaves often withers before flowering time. The flowers, with pink buds and blue flowers, have their stamens sticking out beyond the open corolla tube. The very hairy calyxes do not drop off the fruits. Bees, butterflies, and other insects visit the flowers for nectar. This is a locally common plant of chalky or light dry soils in the south, but is rare in the north. (*June—September*).

5 **Borage** (*Borago officinalis*). The large anthers of this plant form a conspicuous cone in the centre of the flower. The leaves are covered with white bristly hairs and the lower leaves have long stalks. Young sprigs of Borage are sometimes added to salads or wine cups. It is an escape from gardens and is found on waste ground, usually near houses. (*May—August*).

Madwort (*Asperugo procumbens*). A rare plant which has naturalized in a few places by waysides and on waste ground. The bristly stems grow along the ground, and the almost opposite leaves are rough and hairy. The tiny flowers in groups in the leaf axils are pink, then blue (Fig. 1). The calyx develops to form a 2-lipped structure surrounding the fruit. (*May—August*).

Fig. 1

6 **Bugloss** (*Lycopsis arvensis*). The name 'Bugloss' comes from a Greek word meaning 'ox-tongued', which refers to the shape and roughness of the bristly leaves. The plant, about 1 foot to 18 inches high, has long, stalked lower leaves and clusters of almost unstalked flowers, with sepals joined only at the base. The corolla tube is slightly curved and has white scales at its mouth. Bugloss is common on sand-dunes, fields, and waste places throughout Britain. (*May—September*).

1 SEA LUNGWORT   2 ALKANET   3 BLUE GROMWELL
4 VIPER'S BUGLOSS   5 BORAGE   6 BUGLOSS

LIFE SIZE

BORAGINACEAE: FORGET-ME-NOT FAMILY (See p. 215)

1 **Water Forget-me-not** (*Myosotis palustris*). A perennial, with the stem often growing along the ground and sending out runners. The hairs on the leaves are very short, and the calyx hairs are pressed against the surface. The calyx teeth are short and triangular (Fig. 1*a*). The 5 stamens alternate with the 5 blue, rarely white or pink, corolla lobes. The fruit consists of 4 shiny black nutlets. This species is common in wet, shady places throughout the British Isles. (*May—September*).

Three other Forget-me-nots of wet places also have hairs pressed closely against the calyx. Creeping Forget-me-not or Creeping Scorpion-grass (*M. secunda*) has bracts at the base of the lowest flower stalks, which are much longer than the deeply-lobed calyx (Fig. 1*b*) and bend downwards after flowering. It spreads by leafy runners and is fairly common on acid, peaty soils.

Tufted Scorpion-grass (*M. caespitosa*), often found on wet chalky soils, has no runners and has small flowers with the calyx teeth at least half as long as the rest of the calyx.

The rare *M. brevifolia*, of wet places on mountains in the north, has short, broad leaves and pale blue flowers. The stem sends out runners.

2 **Wood Forget-me-not** (*M. sylvatica*). A hairy perennial, with a rosette of leaves at the base, and stems from 6 to 18 inches high. There are bent or spreading hairs on the short, deeply-lobed calyx (Fig. 1*c*). The flowers are comparatively large and bright blue or, rarely, white. This is a rather uncommon, local plant of woods. (*May—July*).

A rare, mountain Forget-me-not, *M. alpestris*, is smaller, with deeper blue flowers and long-stalked hairy leaves from the base of the plant.

3 **Changing Forget-me-not** (*M. discolor*). An annual, about 9 inches high, with slender, hairy stems and leaves and hooked hairs on the calyx. The calyx teeth close up over the fruit as it develops (Fig. 2*a*). The small flowers, which have long corolla tubes, are at first yellow and turn to blue. This is a common plant of meadows and hedges. (*April—September*).

4 **Common Forget-me-not** or Field Scorpion-grass (*M. arvensis*). This species, like Wood Forget-me-not, has a rosette of leaves at the base and bent or hooked hairs on the calyxes. But it has smaller flowers, with saucer-shaped corolla lobes and corolla tubes shorter than the calyx. The name Scorpion-grass refers to the shape of the flowering stem which, like that of many members of the Forget-me-not family, is at first curled up like a scorpion's tail and gradually uncurls as the flowers open. This annual is very common in hedges, woods, and cultivated ground throughout Britain. (*April—September*).

5 **Early Forget-me-not** (*M. hispida*). A small hairy annual, rather similar to Changing Forget-me-not but shorter, with bright blue, rarely white, flowers and a corolla tube shorter than the calyx. The calyx teeth become spread out as the fruits ripen (Fig. 2*b*). It is common on walls and dry banks in south and east England. (*April—June*).

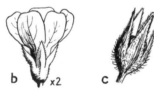

Fig. 1

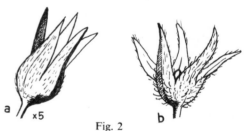

Fig. 2

1 WATER FORGET-ME-NOT   2 WOOD FORGET-ME-NOT   3 CHANGING FORGET-ME-NOT

4 COMMON FORGET-ME-NOT   5 EARLY FORGET-ME-NOT

LIFE SIZE

1 Field Speedwell (Large) or Buxbaum's Speedwell (*Veronica persica*). An annual, sometimes up to 1 foot high, with branched, hairy stems, nearly oval leaves, and flowers borne singly on long stalks in the leaf axils. The flower stalks curve downwards after flowering. The 4 sepals are fairly broad, and the capsule lobes are flattened. As with all Speedwells, there are 4 almost similar corolla lobes and 2 stamens. This species is very common in fields and hedgerows throughout the British Isles. (*February—November*).

Another common Field Speedwell, *V. agrestis*, has smaller, paler flowers on stalks no longer than those of the leaves, and swollen capsule lobes. Grey Field Speedwell (*V. polita*) has darker, grey-green leaves and small, bright blue flowers with pointed sepals.

Ivy-leaved Speedwell (*V. hederifolia*). This slightly hairy annual branches near the base and spreads some distance along the ground. The 3- or 5-lobed Ivy-like leaves have short stalks, and the white or pale lilac petals are no longer than the heart-shaped sepals (Fig. 1). Ivy-leaved Speedwell is common in fields and hedgerows throughout Britain and, indeed, across much of the world. It can be a tiresome weed. (*March—May*).

x ½

Fig. I

2 Wall Speedwell (*V. arvensis*). A small, spreading, hairy annual, less than a foot high, with many flowers on short stalks in leafy spikes. The bracts below the upper flowers are unstalked and untoothed, but the lower leaves have very short stalks and are toothed. This species is common on walls and dry sandy places. (*March—September*).

3 Thyme-leaved Speedwell (*V. serpyllifolia*). A perennial with creeping stems and tufts of upright stems. It has smooth, 3-veined, untoothed leaves and spikes of pale blue or white, short-stalked flowers. It is common in fields and moist waste places throughout Britain. (*March—September*).

The rare Alpine Speedwell (*V. alpina*) of Scottish mountains is a rather fleshy plant with a short spike of small, pale blue flowers. The even rarer Rock Speedwell (*V. fruticans*) has a woody stem and a short spike of fairly large, intensely blue flowers with reddish centres. (*July—August*).

4 Common Speedwell (*V. officinalis*). A small, hairy, much branched perennial with creeping and rooting stems. The flowers, often pale blue or almost pinkish, are on very short stalks in crowded erect spikes in the leaf axils. This is a common plant of dry grassland and heaths, but not on chalky soil. (*May—August*).

5 Brooklime (*V. beccabunga*). This water Speedwell has smooth, round, rather fleshy opposite leaves, with pairs of flower spikes coming from the leaf axils. The small, deep blue flowers have fairly long flower stalks. Brooklime is common in ponds, streams, and wet places, and the creeping stems root or float in the water. (*May—September*).

6 Germander Speedwell or Bird's-eye (*V. chamaedrys*). A hairy perennial with stems which grow along the ground and then turn upwards, and have 2 opposite lines of white hairs on them. The long-stalked, brilliant blue flowers are in spikes arising from the axils of the toothed leaves. Bird's-eye is common in hedges and fields throughout the British Isles. (*March—July*).

Wood or Mountain Speedwell (*V. montana*) is a similar plant, but with stalked leaves, hairs all round the stem, and fewer flowers in each spike. It is a fairly common plant of damp woods. (*April—July*).

7 Water Speedwell (*V. anagallis-aquatica*). The smooth, rather fleshy stems grow up to 1 foot high and carry unstalked leaves, longer than those of other Speedwells. The flower spikes are arranged in pairs, from the axils of 2 opposite leaves. This species is fairly common in wet places throughout Britain. (*June—August*).

Marsh Speedwell (*V. scutellata*), which has white, pink, or pale blue flowers, is shown on page 93.

1 FIELD SPEEDWELL   2 WALL SPEEDWELL   3 THYME-LEAVED SPEEDWELL

4 COMMON SPEEDWELL   5 BROOKLIME   6 GERMANDER SPEEDWELL

7 WATER SPEEDWELL

LIFE SIZE

1 **Meadow Clary** or Meadow Sage (*Salvia pratensis*, family *Labiatae*). A hairy perennial up to 2 feet or more tall, and, like most members of the Deadnettle family, with square stems. It has a 2-lipped corolla with only 2 stamens. The hairy calyx is divided into an upper lip with 3 teeth and a lower lip with 2. The opposite, toothed leaves are wrinkled, and those on the stem are small and unstalked. It is a rare local plant of chalk grassland in southern and central England. (*June—August*).

2 **Common Skullcap** (*Scutellaria galericulata*). This plant and Lesser Skullcap (*see* p. 145) have a small, leafy flap projecting from the upper lip of the calyx. The upright, rather hairy stems grow 6 to 18 inches high from a creeping rhizome. The flowers are usually mostly on one side of the stem. The 4 stamens are hidden within the upper lip of the softly hairy corolla. Skullcap is common in wet places throughout the British Isles. (*June—September*).

3 **Wild Clary** (*Salvia horminoides*). The flowers are often small, with the corolla shorter than the calyx and unopened. The calyx has long white hairs, and there are 2 white spots at the bottom of the lower corolla lip. This plant is fairly common in waste places and by roadsides in the south, becoming rarer further north. (*May—September*).

4 **Corn Salad** or Lamb's Lettuce (*Valerianella locusta*, family *Valerianaceae*). A small annual, with the stem forked several times and blue or pale lilac flowers in small heads at the ends of the branches. The calyx has 5 tiny teeth, and 3 stamens stick out beyond the 5-lobed corolla. The fruits are oval, with a ridge down one side (Fig. 1*a*). Corn Salad is a common plant of fields and hedges, especially on dry soil. (*April—July*).

Keeled Corn Salad (*V. carinata*) is similar, but with lilac flowers and oblong fruit hollowed out on one side (Fig. 1*b*). This is a rare species of south and east England. In two species, *V. dentata* and *V. rimosa*, the calyx forms a tooth-like structure on top of the fruit. The first of these is fairly common in cornfields and has lilac or pinkish flowers and narrow fruits (Fig. 1*c*). The second has pale blue flowers and broad fruits (Fig. 1*d*), and is found occasionally in the south and east.

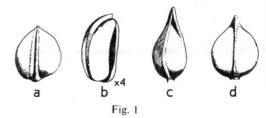

Fig. 1

5 **Spring Gentian** (*Gentiana verna*, family *Gentianaceae*). A small perennial, only a few inches high, with unstalked or very short-stalked flowers arising from rosettes of leaves. The beautiful, bright blue corolla has tiny teeth between the 5 outspread lobes. This Gentian is a rare plant of mountain pastures on limestone in the north and in Ireland. (*April—June*).

6 **Marsh Gentian** (*G. pneumonanthe*). A perennial, with stiff stems up to about a foot in height, and narrow leaves. The corolla tube is greenish-blue outside and clear blue inside. There are tiny teeth between the corolla lobes, which do not open out widely. It is a rare and local plant of bogs and wet heaths in England and Wales. (*August—September*).

1 MEADOW CLARY   2 COMMON SKULLCAP   3 WILD CLARY
4 CORN SALAD   5 SPRING GENTIAN   6 MARSH GENTIAN

LIFE SIZE

1 Cornflower or Bluebottle (*Centaurea cyanus*, family *Compositae*). A slender annual from 1 to 2 or more feet tall, with woolly hairs on the stems and leaves. The lower leaves are toothed or lobed and stalked. The bracts round the flower-heads have narrow, pointed, whitish teeth. The bright blue, sterile, outer florets are larger than the fertile purplish inner ones. The achenes carry a pappus of short hairs. This was once a common cornfield weed but is now rather rare. (*June—August*).

2 Round-headed Rampion (*Phyteuma tenerum*, family *Campanulaceae*). An almost hairless perennial, varying in height from 2 or 3 to about 18 inches. There are short, narrow bracts below the heads of flowers. The 5 long, slender lobes of each corolla form a curved tube in bud, and then gradually spread outwards. There are 5 stamens and usually 2 narrow stigmas. This is a rare, local plant of chalk downs in S. England. (*July—August*).

3 Sheep's-bit or Sheep's Scabious (*Jasione montana*). A hairy plant, branching at the base, with small, short-stalked lower leaves. The bracts below the flower-heads are broader than those of Round-headed Rampion, and the lobes of the straight corolla soon open. The anthers of the 5 stamens are joined together at the base, and there are 2 short stigmas. Sheep's-bit, which in spite of its appearance is not a Scabious, is locally fairly common on dry banks and heaths, except on chalky soils. (*May—September*).

4 Devil's-bit Scabious (*Succisa pratensis*, family *Dipsacaceae*). A rather hairy perennial up to 3 feet high. It differs from Field Scabious and Small Scabious (*see* p. 157) in having almost untoothed leaves and florets all the same length. The lowest leaves are long and rather narrow, and up the stem are pairs of smaller leaves. The flower-heads are surrounded by 2 or 3 rows of hairy bracts, and at the base of each individual floret is a bract as well as a toothed calyx and an extra 4-toothed calyx-like cup (Fig. 1a). Butterflies and bees visit the flowers for the nectar at the bottom of the corolla-tube. The plant has a very short rootstock, said to have been bitten off by the devil — hence its popular name. This is a common plant of meadows, marshes, and damp woods throughout Britain. (*July—October*).

5 Sea Holly (*Eryngium maritimum*, family *Umbelliferae*). The rounded heads of flowers surrounded by prickly bracts look much like members of the Daisy family (*see* p. 218), but an inspection of a single flower (Fig. 1b) reveals 5 pointed sepals, 5 notched petals, and 5 free stamens, characteristic of the Parsley family. The lowest leaves are broad, prickly, 3-lobed, and, like the rest of the plant, covered with a bloom. Sea Holly grows on sandy or shingly places round most parts of the British coasts. (*July—August*).

A rare plant of waste places in S. England, *E. campestre*, has pale green, deeply-cut lower leaves and almost untoothed bracts below the heads of usually whitish flowers.

Fig. 1

6 Chicory (*Cichorium intybus*, family *Compositae*). This plant grows up to 3 feet or more tall, with tough, furrowed stems and long, lobed basal leaves, which are rather hairy. The achenes have a ring of tiny scales instead of the hairy pappus usual in this Dandelion-like group. The roots are used, roasted and ground, to add to some kinds of coffee. The plant is locally common in fields and by roadsides in England and Wales, especially on chalky soil, but rare in Scotland. (*July—September*).

1 Cornflower   2 Round-headed Rampion   3 Sheep's-bit

4 Devil's-bit Scabious   5 Sea Holly   6 Chicory

LIFE SIZE

1 Blackthorn or Sloe (*Prunus spinosa*, family *Rosaceae*). The flowers usually appear early in the spring before the leaves, singly or in pairs on the blackish, spiny branches. There are 5 sepals and petals, numerous stamens, and a single style. The fruit, which has a bluish bloom, has so sharp a taste that it dries up the inside of the mouth when eaten. Country people sometimes make jam or sloe gin from the fruits. The cultivated Plum, Damson, and Greengage are closely related to Blackthorn, which is a common shrub of hedges and woods throughout the British Isles. (*March—May*).

The Bullace or Wild Plum (*P. domestica*) is often found growing wild. It has few spines, usually pairs of flowers appearing with the leaves, and large egg-shaped fruits.

2 Crab Apple (*Malus sylvestris*). A small tree with scaly bark. The flowers are in clusters of about 5 to 7, each with 5 sepals, hairy inside, 5 petals, numerous stamens, and 5 styles, joined together near the base. The yellowish-green, very sour, apple-like fruit (called a pome) often turns scarlet as it ripens. It is used to make crab-apple jelly in country districts. The tree is quite common in woods and hedges in most parts of Britain, rarer in Scotland. (*May—June*).

3 Hawthorn or May (*Crataegus monogyna*). A shrub or small tree with spiny branches, often planted for making farm hedges. The flowers, which have a strong sweet scent, are in clusters at the ends of the short leafy shoots. There are many stamens and a single style with a knob-shaped stigma. The berry-like fruit, the 'haw', contains one stone. This plant is very common in hedges, scrub, and woods throughout the British Isles, except in Scotland. (*May—June*).

The Midland Hawthorn (*C. oxyacanthoides*) is a rarer species, most frequent on clay soils in the east. It has less deeply cut leaves with more rounded lobes, and fewer flowers, usually with 2 styles (Fig. 1).

4 Spindle (*Euonymus europaeus*, family *Celastraceae*). A much-branched shrub or small tree, with green, 4-sided twigs. The flowers may be like those shown in the illustration opposite, with both stamens and ovary, or there may be separate male and female flowers. The flowers are pollinated by insects seeking the nectar round the base of the style. In the autumn the red, fleshy, 4-lobed capsules can be seen splitting open to expose the orange structures (called arils) covering the seeds. The plant is poisonous. Spindle wood is very hard and was at one time used to make skewers and spindles — hence the name. The Spindle is one of the characteristic shrubs of chalky soil, where it is common in woods and hedges, except in the north of Scotland. (*May—June*).

5 Cherry-plum (*Prunus cerasifera*, family *Rosaceae*). A shrub or small tree with greenish twigs, usually without spines. Both twigs and leaves are usually smooth and rather shiny, the leaves appearing at about the same time as the flowers. The yellow and reddish fruit is not formed every year. Cherry-plum is a native of Russia and Persia, and was introduced into this country, where it has become fairly common in hedgerows. (*March—April*).

x½

Fig. 1

LIFE SIZE

1 BLACKTHORN   2 CRAB APPLE   3 HAWTHORN
4 SPINDLE   5 CHERRY-PLUM

1 **Common Buckthorn** (*Rhamnus cathartica*, family *Rhamnaceae*). A shrub or small tree with spreading branches often ending in a thorn. The almost opposite leaves have tiny teeth round the edges. Male and female flowers are developed on separate plants. The male flowers (lower picture) have a 4-lobed calyx and 4 inconspicuous petals opposite the 4 stamens, while in the female flowers (upper picture) there is an ovary with a 3- or 4-lobed style. The black fruits, if eaten, are strongly purgative. In spite of its name, this Buckthorn is a rather uncommon shrub of woods, scrub, and hedges on chalky soil in England, Wales, and parts of Ireland. (*May—June*).

woods, hedges, and scrub throughout most of Britain. (*May—August*).

Fig. 1

2 **Alder Buckthorn** (*Frangula alnus*). A rather slender thornless shrub or small tree with alternately arranged, untoothed leaves and male (stamens) and female (ovary) parts in the same flower. The parts of the flower are in fives, and the fruits turn from red to dark purple. The leaves usually have 6 or 7 pairs of clearly marked veins — more than do those of Common Buckthorn. Alder Buckthorn is locally common in scrub, hedges, and at the edge of bogs in England and Wales, rare in Ireland. (*May—July*).

4 **Sea Buckthorn** (*Hippophaë rhamnoides*, family *Elaeagnaceae*). A shrub or small tree with side shoots often ending in a thorn. The alternately arranged leaves are covered with silvery scales, especially on the under side. The tiny greenish flowers appear before the leaves, and male and female ones are on separate plants. The petal-less female flower consists of a small 2-lobed calyx surrounding the ovary, with a single style. The male flower has 2 sepals and 4 stamens, the pollen from which is carried to the female plants by the wind. Sea Buckthorn grows on sand-dunes and sea cliffs, mainly along the east coast, but has been planted elsewhere. (*March—May*).

3 **Holly** (*Ilex aquifolium*, family *Aquifoliaceae*). This evergreen shrub or small tree, with its characteristic tough, glossy, and prickly leaves, is the only British member of its family. The twigs are very dark green, and the upper leaves are often unlobed, with a prickly point at the end only. The parts of the flowers are in fours (Fig. 1), the male and female flowers being often on separate trees. The red berries, which develop from the female flowers only when a male tree is nearby, ripen in September and usually hang on through the winter unless eaten by birds. Holly is common in

5 **Privet** (*Ligustrum vulgare*, family *Oleaceae*). An almost evergreen shrub, of the same family as the Ash (*see* p. 197). It has slender branches and very short hairs on the young shoots. The opposite, untoothed, shiny leaves are narrower than those of the species most frequently planted as a garden hedge. The masses of flowers have a strong, rather unpleasant scent. Each flower has a small, cup-shaped calyx, a 4-lobed corolla, 2 stamens, and a single style. Privet is common in woods, hedges, and scrub in S. England, where it is one of the characteristic shrubs of chalky soils. (*June—July*).

1 COMMON BUCKTHORN   2 ALDER BUCKTHORN   3 HOLLY
4 SEA BUCKTHORN   5 PRIVET

LIFE SIZE

1 Larch (*Larix decidua*, family *Pinaceae*). This is the only deciduous British conifer, that is, the only conifer to lose all its leaves in the autumn. In the spring, this attractive tree produces new needle-like leaves in tufts on very short shoots on the longer, often drooping twigs. The flowers are cones, as in all members of the Pine family. The female cones are crimson, with the ovules borne on scales, and the male cones are yellow and bear the pollen sacs. As the winged seeds ripen, the female cones turn brown. This tree is often planted in gardens and parks and for timber. (*March—May*).

2 Scots Pine or Scots Fir (*Pinus sylvestris*). This tall tree is the typical 'fir' shape when young, though older trees often have a spreading crown of branches. The bark of mature trees is reddish, with bare patches where branches have fallen off. The twisted, needle-like leaves are in pairs on short shoots along the twigs. There are separate male and female cones on the same tree. The cones take about 2 years to ripen, and then in dry weather the scales of the cone open to release the winged seeds. Scots Pine, which is the only true native Pine, is common in Scotland and parts of Southern England and is frequently planted in other parts of Britain. (*May—June*).

3 Yew (*Taxus baccata*, family *Taxaceae*). An evergreen tree, with the flattened 'needles' arranged in 2 opposite rows along the twigs. The inconspicuous male (upper picture) and female flowers are on separate trees. Birds eat the red, waxy cup (aril) round the seed, but they do not digest the seed, which is poisonous, as are the leaves. Yew is fairly common on chalky soil, especially in S.E. England. It used to be often planted, especially in country churchyards, for the wood was prized for making bows. (*March—April*).

4 Juniper (*Juniperus communis*, family *Cupressaceae*). An evergreen shrub or small tree, with prickly leaves in whorls of 3. The tiny male and female cones are on separate plants. The male consists of several scales to which the pollen sacs are attached, and the female of 3 fleshy scales joined together to make a berry-like structure, which turns bluish-black. These 'berries' are used to flavour gin. Juniper is locally common on chalk downs and scrub, on heaths, and in woods, scattered throughout Britain. (*May—June*).

5 Spruce (*Picea abies*, family *Pinaceae*). A tall, evergreen, 'fir'-shaped tree, which is often called Norway Spruce as it is a native of Scandinavia. The branches are in whorls up the trunk and bear 4-sided 'needles' with sharp-pointed tips, which are attached to little pegs on the twigs (Fig. 1*a*). These pegs are arranged round the twig in a spiral. The young female cones, at the ends of branches, are green or purple, and turn brown; the male ones are yellow or red. This tree is often planted for timber in Britain, and the young plants are the Christmas trees. (*May—June*).

Silver Fir (*Abies alba*). This tall evergreen has a grey bark and needles spreading out on two sides of the branch, instead of all round it as in the Spruce. The needles have two white lines along the under side and leave a circular scar on the twig when they fall off (Fig. 1*b*). The ripe cones are upright. This tree is now seldom grown for timber because it easily becomes diseased, and so is seen mainly in parks and gardens. (*May—June*).

Douglas Fir (*Pseudotsuga taxifolia*). This is much like Silver Fir, but has needles green on both sides which leave sloping scars on the twig (Fig. 1*c*). The slender, pointed buds are a bright brown colour, and the ripe cones hang downwards. This common North American tree has been planted in Britain, and has occasionally naturalized. (*May—June*).

a

b

c

Fig. 1

1 LARCH   2 SCOTS PINE   3 YEW
4 JUNIPER   5 SPRUCE

LIFE SIZE

SALICACEAE: WILLOW FAMILY (See p. 214)

Willows (*Salix*). Trees or shrubs with alternately arranged leaves (but *see* No. 2) and erect male and female catkins on separate trees. The male catkin consists of numerous bracts, each with 2 or more stamens in its axil, and the female catkin has an ovary in the axil of each of the many bracts. The fruit is a capsule splitting into 2 parts, which curl back to release the hairy seeds (Fig. 1).

There are many British Willows, with crosses between them, so that they are difficult to name, especially as the catkins usually appear while the leaves are very young. A few are described here, and some more on page 199.

Fig. 1

1 Great Sallow or Pussy Willow (*Salix caprea*). A tall shrub or small tree, which flowers before the leaves develop. The young twigs and buds are covered with tiny hairs, but the more mature twigs are smooth and often reddish. The fully-grown leaves have a greyish felt of hairs on the under side. The large male catkins (left picture) have long silky hairs on the bracts and are much more conspicuous than the female (right picture). This is a common plant of scrub, woods, and hedges throughout Britain. It is used to make hurdles and fences. (*March—April*).

Eared Sallow (*S. aurita*). A common and widespread bush with spreading branches and smaller in all its parts than Great Sallow and Common Sallow. The leaves, which have large leafy stipules at the base (Fig. 2), are wrinkled, and soft and hairy on the under side. The twigs are ridged like those of Common Sallow. (*April—May*).

Tea-leaved Willow (*S. phylicifolia*) is an erect shrub growing along riverbanks and in wet, rocky places, mostly in the north. It has shiny, reddish-brown twigs and smooth leaves, green and glossy on top, which usually appear with the catkins. (*March—May*).

Fig. 2

x½

Common Sallow or Grey Willow (*S. atrocinerea*). Similar to Great Sallow but with smaller, narrower, almost untoothed leaves and very hairy buds and

young twigs. Ridges can be seen on the older twigs if the bark is peeled off. This Willow is common in woods, marshes, heaths, and by streams throughout Britain. (*March—April*).

2 Purple Willow (*S. purpurea*). A bush with smooth buds and twigs. Unlike those of other Willows, the leaves are often opposite, and when fully developed are often dull bluish-green above and have a whitish bloom below. The stems, which become purple when boiled, contain a substance called salicin in the bark which was much used in medicines before the discovery of aspirin. Purple Willow is locally common in fens, marshes, and on riverbanks, mainly in England. It is sometimes planted as an osier (*see* No. 3). (*March—May*).

3 Common Osier (*S. viminalis*). A shrub with long, flexible branches, much used for basket-work, for which it is often grown. The long, narrow leaves have soft, silvery hairs on the under side. This Willow is common by streams and in wet places throughout Britain, except in hilly districts. (*March—June*).

4 White Poplar (*Populus alba*). A fairly tall tree with smooth grey bark and white, hairy buds and young twigs (*see* pp. 204-5). The toothed leaves, which vary from almost unlobed to strongly 5-lobed, have a white-felted under side. The long, drooping male catkins have numerous stamens to each hairy bract. The female ones are much shorter. The capsules and seeds are like those of Willows. This is a rather uncommon wild tree of damp woods, mainly in S. England, but is often planted. (*March—April*).

The taller and more common Grey Poplar (*P. canescens*) has usually unlobed leaves with a few large teeth and grey under sides. The trees often produce suckers, which sometimes have leaves like those of White Poplar.

5 Black Poplar (*P. nigra*). A tall tree with black, furrowed bark and branches gradually bending downwards (*see* pp. 204-5). The leaves are green on both sides, with flattened stalks. It occurs in wet woods in the midlands and east, and has been planted elsewhere. (*March—April*).

The more common Black Italian Poplar (*P. x canadensis* var. *serotina*) has branches growing upwards to form a fan-shaped crown. Lombardy Poplar (*P. italica*), which is not a native tree, but is very often planted, has a characteristic tall, narrow shape.

LIFE SIZE

1 Pussy Willow   2 Purple Willow   3 Common Osier
4 White Poplar   5 Black Poplar

1 Common Lime (*Tilia x vulgaris*, family *Tiliaceae*). This large, commonly planted tree (*see* pp. 202-3) is a cross between Small-leaved Lime and Large-leaved Lime. The young twigs are smooth, and the thin leaves are up to 2 or 3 inches long, with tufts of white hairs underneath in the angles between the veins. The flowers are in hanging clusters of 4 or more with a long bract joined to the stalk. They are sweetly scented and visited by bees for nectar. (*July*).

Large-leaved Lime (*T. platyphylla*) differs from Common Lime, not only in the larger leaves, but also because there are hairs on the young twigs and all over the under side of the leaves. The flowers, usually in threes, appear about the end of June. Lime flowers were at one time used to make a delicately flavoured tea.

2 Small-leaved Lime (*T. cordata*). The small leaves, usually 1½ to 2½ inches long, have tufts of brownish hairs underneath. The flower clusters do not hang downwards, and the fruits are much less ribbed than those of the other Limes. This tree is found wild in parts of England and Wales, especially in Ash woods, and it has also been planted. (*July*).

3 Hornbeam (*Carpinus betulus*, family *Corylaceae*). A small tree with a grooved trunk and smooth grey bark. The almost hairless leaves have parallel veins on either side of the midrib. The female flowers are in catkins hanging downwards at the end of the twigs, and the male catkins arise from buds on the previous year's twigs. Each fruit is attached to a leafy, 3-lobed bract. Hornbeam is a native tree of S.E. England and has been planted elsewhere. (*March—May*).

4 Hazel (*Corylus avellana*). A large shrub with smooth brown stems, often found in Oak woods and hedges, where it is cut back periodically. The young twigs, leaves, and leaf stalks are somewhat hairy. The male catkins (lambs' tails) are long and drooping, yellow with pollen in the early spring before the leaves appear. The female flowers are small, upright, and hidden by bracts except for the long red styles. The edible nut is surrounded by a green leafy cup, and ripens in late September or October. Hazel is common throughout most of Britain. The stems are used for making baskets, hurdles, crates, etc. (*February—April*).

5 Horse Chestnut (*Aesculus hippocastanum*, family *Hippocastanaceae*). This large tree (*see* pp. 204-5) is easily recognized in winter and spring by the horse-shoe-shaped leaf scars on the twigs (Fig. 1) and the sticky brown buds, in early summer by the 'candles' of white flowers, and in autumn by the large smooth brown seed in a prickly capsule. The seeds, called 'conkers' by children, are used for a game originally played with snail shells ('conkers' in dialect). A 'conker' is threaded on to a string and hit against that of an opponent. The fruit of Horse Chestnut is rather like that of Sweet or Spanish Chestnut (*see* p. 200), but the trees are, in fact, not related. (*May—June*).

Fig. 1

ONE QUARTER LIFE SIZE
WITH LIFE SIZE DETAILS

1 COMMON LIME	2 SMALL-LEAVED LIME	3 HORNBEAM
4 HAZEL	5 HORSE CHESTNUT	

1 **Bog Myrtle** or Sweet Gale (*Myrica gale*, family *Myricaceae*). A small shrub, 2 to 4 feet high, with reddish stems. The flowers appear before the leaves, male and female catkins being usually produced on separate plants. The male catkins (left-hand picture) consist of numerous broad, untoothed bracts, each with 4 to 18 short stamens in its axil. The small female catkins (right-hand picture) have an ovary with 2 red styles in the axil of each bract. The fruit is small, with 2 wings. Yellow dots (glands) are scattered over the leaves, which have a pleasant aromatic scent. Bog Myrtle is locally common in fens and bogs throughout most of Britain. (*April—June*).

2 **Barberry** (*Berberis vulgaris*, family *Berberidaceae*). A shrub with ridged twigs bearing 3-pronged spines, which are modified leaves. There are short shoots with a tuft of ordinary leaves in the axil of each spine, and drooping groups of flowers at the ends of the shoots. The flowers consist of several whorls of more or less similar perianth segments, 6 stamens, and a single carpel. The fruit is a red berry. Barberry is a widespread but rather rare plant of hedges and woods. (*May—June*).

The related Oregon Grape (*Mahonia aquifolium*), a North American species often planted in Britain, has no spines. The evergreen leaves are divided into leaflets, and the berries are blackish.

3 **Tamarisk** (*Tamarix gallica*, family *Tamaricaceae*). A partially evergreen shrub with slender branches and tiny, pointed, scale-like leaves with a bloom on the surface. The pink or white flowers are in crowded spikes, each flower having 5 sepals, petals, and stamens, and an ovary with 3 styles. The fruit is a capsule splitting into 3 to release the seeds, each of which is topped by a tuft of hairs. Tamarisk is a frequent hedge-plant by the sea in the south. (*July—October*).

4 **Spurge Laurel** (*Daphne laureola*, family *Thymeleaceae*). This small evergreen shrub, 1 to 3 feet high, is neither a Spurge nor a Laurel. It has tough, leathery leaves, mostly near the tops of the stems. The clusters of petal-less flowers, which grow in the leaf axils, have 4 sepals, 8 stamens, and an ovary with a large, rounded stigma. The fruits are poisonous and black when ripe. This is a fairly common plant of woods and shady places on chalky soils in England and parts of Wales. (*January—April*).

The rare Mezereon (*D. mezereum*) is found occasionally in woods on chalky soils. The fragrant flowers are pink and appear before the light green leaves. The poisonous fruits are red when ripe. (*February—April*).

Butcher's Broom (*Ruscus aculeatus*, family *Liliaceae*). A stiff shrub, up to 2 feet or more high, with green stems bearing what appear to be thick, smooth green leaves with prickly points at the end. These are really modified branches, and the tiny greenish-white flowers are developed on them (Fig. 1). The male and female flowers are on different plants, the male having 3 stamens. The fruit is a red berry. This is a local plant of hedges and woods in the south, though often planted elsewhere. (*January—May*).

$\times \frac{1}{2}$

Fig. 1

1 Bog Myrtle   2 Barberry   3 Tamarisk
4 Spurge Laurel

LIFE SIZE

1 Guelder Rose (*Viburnum opulus*, family *Caprifoliaceae*). A shrub or small tree with 3- or 5-lobed opposite leaves, which turn red in autumn. There are small stipules at the base of the leaf stalks and pimples on the twigs. In early summer the 'flower-heads' consist of a ring of large white sterile flowers round a group of scented inner fertile ones. The latter have 5-lobed calyxes and corollas, 5 stamens, and 3 stigmas. The bright red, berry-like fruits each contain one seed. Guelder Rose is fairly common in hedges and open woods, except in Scotland. (*May—July*).

2 Dogwood (*Cornus sanguinea*, family *Cornaceae*). The straight, red stems of this shrub are very conspicuous in hedges and woods in winter. It usually grows to about 6 feet high and has dark-green, untoothed, opposite leaves that are slightly hairy and turn dark red in autumn. The Hawthorn-scented white flowers each have a small 4-toothed calyx, 4 separate petals alternating with 4 stamens, and a single style. The black fruits are not edible. Dogwood is common in hedges, scrub, and open woods, mainly in southern England where it is one of the characteristic shrubs of chalky soils. (*June—July*)

x 1

Fig. 1

Dwarf Cornel (*Chamaepericlymenum suecicum*). A slender plant, with a few stems and short branches, less than a foot high and not easy to recognize as related to Dogwood. The leaves are unstalked and have only 1 or 2 pairs of side veins. The head of minute purple flowers is surrounded by 4 white, petal-like bracts (Fig. 1). The fruit is red when ripe. This is a rare moorland plant of northern mountains. (*July—September*).

3 Wayfaring Tree (*Viburnum lantana*, family *Caprifoliaceae*). A shrub or small tree with a thick coat of short, white or reddish-brown hairs on the young shoots and the under sides of the soft, wrinkled leaves. The flowers are scented and are like those of the related Guelder Rose, but without the sterile outer flowers. The one-seeded fruits are black when ripe. Like Dogwood, this is a characteristic shrub of chalky soils, common in hedges and open woods in the south, but rare in the north. (*May—June*).

4 Elder (*Sambucus nigra*). An unpleasant-smelling small tree or shrub with a soft, white pith inside the stems. The opposite leaves usually consist of 5 almost hairless leaflets, and the creamy-white flowers are in flat-topped masses. Each flower has a 5-lobed calyx and corolla, 5 stamens attached to the corolla, and 3 or 5 stigmas. The black fruits are sometimes used to make jelly or Elderberry wine. Elder flower water, made from the flowers, used to be thought good for the complexion. It is a common plant of woods, hedges, and waste ground, especially on chalky soils, although rare in the north of Scotland. (*June—July*).

Danewort or Dwarf Elder (*S. ebulus*). This perennial, up to 3 or 4 feet high, has also a strong, unpleasant scent. The stems, which die down in winter, grow up in the spring from a creeping rhizome and carry pairs of leaves, each divided into 7 or more leaflets, often longer and narrower than those of Elder. There is a pair of tiny leaflets at the base of the leaf stalk. The scented white flowers are sometimes tinged with purple, and the fruit, when ripe, is black. This is a rather uncommon plant of hedges and roadsides, mostly in the south. (*July—August*).

HALF LIFE SIZE

1 GUELDER ROSE   2 DOGWOOD   3 WAYFARING TREE
4 ELDER

ROSACEAE: ROSE FAMILY (See p. 212)

1 **White Beam** (*Sorbus aria* agg.). A small tree or shrub with dark-grey bark and green winter buds. The toothed leaves vary very much in shape, but are usually unlobed and dark green on the upper side, with a white felt of hairs underneath. The fairly large flowers have 5 sepals and petals, many stamens, and 2 styles. The red, spotted fruits are eaten by birds. This is a locally common tree of woods on chalk and limestone. (*May—July*).

There are two similar, but rare, groups of species: *S. intermedia* agg. has leaves lobed, but not deeply, tapering at the base, and grey-felted underneath, and red fruits. *S. latifolia* agg. also has a grey felt beneath the leaves, which have only very small triangular lobes and often a rounded base. The fruits are orange or brownish. Both groups grow in woods in hilly districts, mostly in the west. These species all hybridize freely with each other.

2 **Rowan** or Mountain Ash (*S. aucuparia*). This small tree, with fairly smooth, greyish-brown bark and hairy, brown buds, is not related to the Ash. It differs from both Ash and Elder in having the leaves arranged alternately instead of opposite. The dark green leaves consist of several pairs of toothed leaflets with an odd leaflet at the end. The creamy-white, scented flowers have many stamens and 3 or 4 styles. The brilliant orange-red fruits are eaten by birds. Rowan is fairly common in woods on hills, especially in the north and west. (*May—June*).

3 **Bird Cherry** (*Prunus padus*). A small tree or shrub with brown bark and shiny leaves, which have tufts of hairs underneath in the angles between the side veins and the midrib. The long, loose spikes of white flowers appear with the leaves, often hanging down. Birds eat the bitter, black fruits. This is a rare plant of hedges and woods in N. England and S. Scotland. (*April—May*).

4 **Wild Cherry** or Gean (*P. avium*). A sometimes tall tree, with shiny grey bark which comes off in strips. The pale green leaves have short hairs on the under side, especially in the angles between the veins and the midrib. The clusters of large white flowers come at the same time as the leaves. The sepals bend backwards after the flowers open. The fairly large fruits are red or blackish. This locally common tree grows by roadsides and in oakwoods and beechwoods throughout the British Isles, but is rare in the north of Scotland. (*April—May*).

Sour Cherry (*P. cerasus*) is a rare hedge-plant found mainly in the south on acid soils. It is smaller than Wild Cherry and produces numerous suckers. The glossy, dark green leaves are hairless and stiff when fully developed, and the red fruits are bitter.

5 **Wild Service Tree** (*Sorbus torminalis*). A shrub or fairly tall tree with scaly grey bark and green winter buds. The leaves are deeply cut into pointed lobes and are practically hairless underneath; but the flower stalks are hairy. The numerous, rather large, white flowers have 2 styles that are joined together at the bottom. Birds eat the brown, spotted fruits. This rare tree grows in woods and hedges on clay or limestone in the southern half of England. (*May—June*).

THREE-QUARTERS LIFE SIZE

1 WHITE BEAM   2 ROWAN   3 BIRD CHERRY

4 WILD CHERRY   5 WILD SERVICE TREE

1 **Ash** (*Fraxinus excelsior*, family *Oleaceae*). A tall tree with a grooved, grey trunk (*see* pp. 202-3). It can be recognized in winter by the hard, black buds on smooth, grey twigs and in summer by the long, opposite leaves, divided into several pairs of leaflets. The groups of purplish flowers, each consisting of stamens and/or ovary only, appear before the leaves, are pollinated by the wind, and soon turn greenish. The fruits, called 'keys', have a wing at the end. Ash is common in hedges and woods, especially damp woods on chalky soil in the north and west, and sometimes grows by streams. The hard wood makes useful timber and is a particularly good burning wood. (*April—May*).

2 **Maple** (*Acer campestre*, family *Aceraceae*). A small tree which sometimes grows as a bush in hedges. It has opposite, usually 5-lobed leaves, which are like those of Sycamore, to which it is related, except that they are smaller and the lobes more rounded. The upper surface of the leaves is dark green, and the lower surface paler and somewhat hairy. They turn bright red and yellow in the autumn. The greenish male and female flowers (**2a**) point upwards. They consist of 5 tiny sepals and petals, 8 stamens in the male flowers, and an ovary with 2 styles in the female. The winged fruits are carried away by the wind. Maple is common in hedges and woods, especially on chalky soils, in south England and the Midlands; rare elsewhere in Britain. The stems contain a sugary sap, which in the Sugar Maples of North America is extracted for syrup. The wood is very ornamental and has been used to make bowls, furniture, etc. (*May—June*).

3 **Sycamore** (*A. pseudoplatanus*). A tall tree with smooth, grey bark, peeling off in strips from the older trunks (*see* pp. 204-5). The twigs are brown and the buds green. The opposite, toothed, 5-lobed leaves are green above, shiny and paler below, often on red stalks. The greenish-yellow flowers and the fruits are similar to those of Maple, but the flowers hang downwards and the wings of the fruits are less widely spread. Sycamore, which is not a native plant, has often been planted and has now naturalized throughout Britain, in woods, hedges, and parks. The wood has, among other things, been used as a veneer. (*April—June*).

4 **London Plane** (*Platanus acerifolia*, family *Platanaceae*). This tree is a cross between the Oriental Plane (*P. orientalis*) and the American Plane (*P. occidentalis*). It is not a native of Britain, but it withstands a sooty atmosphere better than most broad-leaved trees, and for that reason it has been widely planted in London and other large towns. The tree can be recognized at all times of the year by the bark, which comes off in small pieces, leaving yellow patches underneath, and by the hanging balls of fruits, which remain on the tree during the winter. The leaves differ from those of Sycamore in being alternately arranged, and in having fewer teeth. The small male or female flowers (**4a**) usually have 8 perianth segments, 4 stamens, and an ovary with a single style. (*May*).

HALF LIFE SIZE

1 ASH   2 MAPLE   3 SYCAMORE

4 LONDON PLANE

1 English Elm (*Ulmus procera*, family *Ulmaceae*). A tall tree with rough, furrowed bark and thick, spreading branches (*see* pp. 202-3). The tree often throws up many suckers at the base. The twigs and leaf stalks are hairy, and the fully developed leaves, which turn yellow in the autumn, have rough hairs above and short, soft ones below. The leaf base is larger on one side than the other. The reddish flower clusters (1a) appear before the leaves, each flower consisting of 4 or 5 perianth segments, the same number of stamens, and an ovary with 2 styles. The fruit is surrounded by a flat, green wing. This is a common tree of roadsides and woods, mainly in S. England and the Midlands. (*February—March*).

2 Wych Elm (*U. glabra*). A tall tree with few suckers, which can be distinguished from English Elm by the larger, very rough leaves, 3 to 6 inches long when fully developed. They have shorter stalks that are often nearly hidden by the rounded base of one side of the leaf. Wych Elm, although it grows throughout Britain, is more often found in the west and north. Hybrids between it and other species are common. (*February—March*).

3 Crack Willow (*Salix fragilis*, family *Salicaceae*). Unlike the Willows described on page 186, Crack Willow, if it has not been pollarded, grows into a fairly large tree. It has a greyish, grooved bark, and the branches grow out at a wide angle to the trunk. The twigs and even the larger branches break off very easily. The long, smooth, green leaves often have a bloom on the paler under side when fully developed, and may have irregularly-shaped tips. The drooping catkins appear with the leaves on short side-shoots, and the male (3a) and female catkins are on separate trees. This is one of the commonest British Willows of riversides and wet woods, except in the far north. (*April*).

White Willow (*S. alba*). A fairly tall tree with more upright branches and twigs which break off much less easily than those of Crack Willow. The leaves, which are rather similar in shape to those of Crack Willow, are more finely toothed, and white underneath with long silky hairs closely pressed to the surface. Like Crack Willow, it grows by rivers and in wet woods throughout Britain, and is often pollarded. One variety is used for making cricket bats. (*April—May*).

Almond Willow (*S. triandra*). This Almond-scented shrub or small tree, with peeling bark and reddish-brown twigs, is common by riversides. The fairly long, finely-toothed, shiny, dark green leaves have small leafy stipules at the base (Fig. 1a). The short, upright catkins appear with the leaves, and the trees often flower a second time later in the summer. The male catkins have 3 stamens to each bract. (*March—May, July—August*).

Bay Willow (*S. pentandra*). A shrub or small tree with shining brown twigs and smooth, Bay-scented, dark green leaves that are broader than those of the other Willows described here (Fig. 1b). The male catkins have 5 stamens to each bract. Bay Willow is fairly common in wet places in the north. (*May—June*).

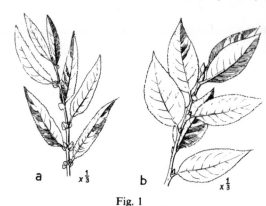

Fig. 1

4 Creeping Willow (*S. repens*). A low, bushy, creeping shrub, with long, white, silky hairs on the young leaves and twigs and on the under sides of the older leaves. The short-stalked catkins appear just before the leaves. It is locally common on heaths and moors throughout the British Isles. (*March—May*).

Least Willow (*S. herbacea*) is a much smaller creeping shrub, often only a few inches high. It has creeping rhizomes and short branches bearing a few small, rounded leaves which are smooth and green on both sides and slightly toothed. It is locally common on mountains in Wales, Ireland, and the north. (*June—July*).

5 Aspen (*Populus tremula*). The broad, green leaves on long, flattened stalks are moved by the gentlest breeze and make a continuous rustling sound. The tree may grow 60 or 70 feet high; it has grey bark and smooth twigs and leaves — although the young leaves and the leaves on suckers are sometimes covered with silky hairs. The drooping male catkins (5a) have reddish anthers, and the female ones purple stigmas. Aspen is common in woods in the north and west, but less common elsewhere. (*February—April*).

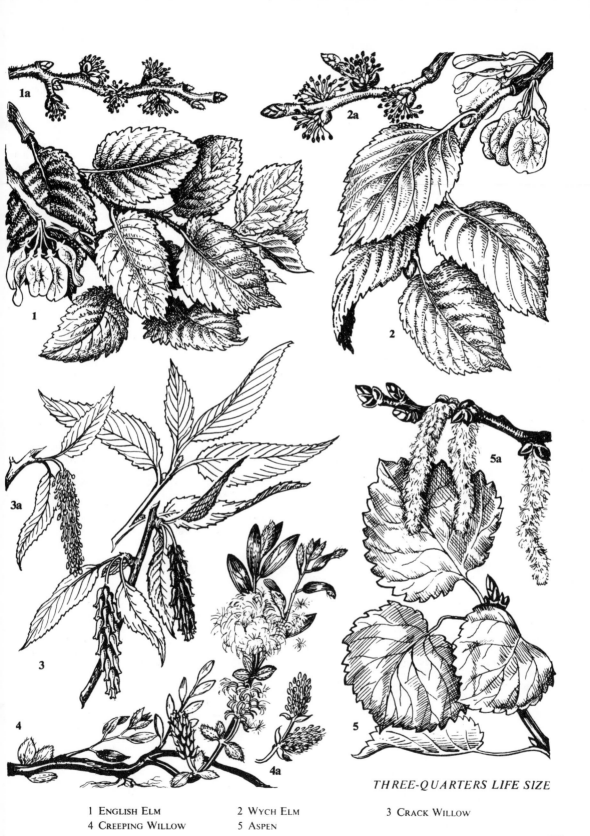

1a
2a
1
2
3a
5a
3
4
4a
5

THREE-QUARTERS LIFE SIZE

1 ENGLISH ELM
4 CREEPING WILLOW

2 WYCH ELM
5 ASPEN

3 CRACK WILLOW

1 **Silver Birch** (*Betula verrucosa*, family *Betulaceae*). A small tree with rough, black bark near the base of the trunk and thin, smooth, white, peeling bark above. The branches are drooping and bear hairless twigs and light green leaves with very pointed tips. The yellowish male catkins appear in the leaf axils in autumn but do not open until the following spring when the female catkins appear. In the male catkins there are 6 stamens, in the female 3 ovaries within each bract. The bracts in the female catkins are 3-lobed. The 2-winged fruit is tiny. Silver Birch is common in woods and heaths on light, but not chalky, soils, especially in the east and south. Birch wood is used for making furniture and the twigs are made into brooms. (*April—May*).

2 **Downy Birch** (*B. pubescens*). The whole trunk is smooth and greyish, rarely white, and the branches are spreading or turned upwards, with very hairy young twigs. The rather dull green leaves have shorter pointed tips than those of Silver Birch, and they are usually hairy underneath. This species is fairly common in woods and heaths, especially in the west and north. The Scottish mountain form is less hairy, but the buds are sticky and the young leaves smell of resin. Crosses between Silver and Downy Birch are quite common. (*April—May*).

Dwarf Birch (*B. nana*). A shrub or dwarf tree with spreading branches that often grow along the ground. The twigs bear tiny, rounded, much-toothed leaves on very short stalks. It is locally common on moors and in bogs in Scotland and is one of a group of plants found both in the arctic and on mountains further south. (*May—June*).

3 **Alder** (*Alnus glutinosa*). This fairly small tree has furrowed brownish-black bark and short branches. The smooth twigs bear stalked buds, and the young leaves are somewhat sticky — hence the Latin name *glutinosa*. The fully developed leaves are dark green, almost hairless, with flat or slightly notched tips. The catkins appear in autumn and develop in spring before the leaves. The drooping male catkins have 12 stamens to a bract and the smaller, sturdy female ones have 5-lobed, black, woody bracts that remain after the seeds have dropped out, and look like tiny

Pine cones. The tree grows in wet fen woods and by streamsides throughout Britain. (*February—April*).

4 **Oak** (*Quercus robur*, family *Fagaceae*). A large tree with rough, furrowed, brown bark, crooked branches, and knobbly twigs (*see* pp. 202-3). The hairless leaves are short-stalked, but the fruits (acorns) are fairly long-stalked. The long, hanging male catkins have lobed perianths and usually 6 stamens per flower. The female flowers are in tiny groups of 1 to 5, each with 3 styles. This is a common tree of hedges and woods throughout Britain, especially on heavy soils in the Midlands and south. The timber used to be used to build ships, for the beams in houses, and for many other purposes, and it is still used for building work and furniture. (*April—May*).

Sessile Oak or Durmast Oak (*Q. petraea*) differs in having longer leaf stalks and leaves hairy beneath, without rounded lobes at the base. The acorns are unstalked or very short-stalked. This species is also common throughout most of Britain, but mainly on acid soils in the north and west, and sandy soils in the south and east.

5 **Beech** (*Fagus sylvatica*). A large tree with smooth grey bark, a straight trunk, and spreading branches (*see* pp. 202-3). The brown winter buds are very long with pointed tips. The beautiful light-green leaves are fringed by long silky hairs when young. The long-stalked male flowers consist of a lobed perianth and numerous stamens; the female ones are surrounded by a cup that develops into a rough, brown husk enclosing the 1 or 2 nuts, called the 'mast'. Beech is a native tree of the south-east and has been planted elsewhere in Britain. It casts a dense shade, and beechwoods on chalk soil have little undergrowth. (*April—May*).

Sweet Chestnut or Spanish Chestnut (*Castanea sativa*). A tall tree, not related to Horse Chestnut (p. 187), with furrowed bark, long, evenly toothed leaves, and erect catkins. The 2 or 3 edible brown nuts are inside a prickly case. This tree is not native to Britain, but is often planted and is common in the south-east. (*June—July*).

HALF LIFE SIZE

1 SILVER BIRCH   2 DOWNY BIRCH   3 ALDER
4 OAK   5 BEECH

OAK

ASH

BIRCH

WYCH ELM

COMMON ELM

BEECH

LIME

WINTER

OAK

ASH

BIRCH

WYCH ELM

COMMON ELM

BEECH

LIME

SUMMER

SYCAMORE

HORSE CHESTNUT

BLACK POPLAR

LARCH

WHITE POPLAR

MOUNTAIN ASH

CRACK WILLOW

WINTER

SYCAMORE

HORSE CHESTNUT

BLACK POPLAR

LARCH

WHITE POPLAR

MOUNTAIN ASH

CRACK WILLOW

SUMMER

NAMING PLANTS

Latin names have been used by scientists for both plants and animals for several hundred years. Latin was used in the first place because in the middle ages it was the international language which was understood by educated people throughout the civilized world. Names in Latin, therefore, could be understood equally well in any country, whereas names in the local language could not. Although fewer people know Latin today, it is still used as an international language among scientists.

The Latin name of a plant often refers to some feature of the plant which distinguishes it from others. For instance, the Latin name for Wood-sorrel (p. 85) is *Oxalis acetosella*: *Oxalis* comes from two Greek words meaning sour and salt, and *acetosella* from the Latin word for vinegar. The three common Plantains (p. 61) are called *Plantago major* — the largest one; *Plantago media* — the medium one; and *Plantago lanceolata* — the Ribwort Plantain, which has long, narrow leaves shaped like a lance.

The Latin name may refer to something else of importance about the plant. For instance, *hirta* or *hirsuta* means that the plant is hairy; *odorata* means a sweet-smelling plant or flower; *alba* or *album*, a white flower; *repens*, a plant which creeps and roots; *procumbens*, a plant with prostrate, non-rooting stems; *parviflora*, small-flowered, and so on. The Latin name *officinalis* comes from the word *officina*, used for the storeroom of a medieval monastery where medicines were kept. It is given to many plants from which drugs can be obtained. Examples are: *Taraxacum officinale*, the Dandelion (p. 41) used for treating gall and liver complaints, and *Saponaria officinalis*, the Soapwort (p. 107), used to treat coughs and asthma. The Latin name may mean the same as the English name: for instance, *Saponaria* comes from the Latin word for soap, and both Latin and English names refer to the fact that the crushed roots froth up with water, like soap, and may actually be used as soap.

Plants are sometimes named after famous botanists who, perhaps, first identified them: for example, *Tofieldia*, a rare Asphodel like a small Bog Asphodel (p. 15), named after the Yorkshire botanist Tofield; or *Lonicera*, Honeysuckle (p. 15), named after the German botanist Lonitzer. Latin names also often indicate where a plant grows. For instance, *Silene maritima* is the Campion which is found near the sea (p. 73); *Ranunculus aquatilis* is the Crowfoot which grows in water (p. 67); and *Myosotis sylvatica* is the Wood Forget-me-not (p. 173). *Arvensis, campestris,* or *pratensis* usually mean plants of grassland or meadows; and *palustris* means a plant of wet places and marshes. Other names are easy to guess: *vulgare* means a common plant, *verna* one that flowers in the spring, and so on.

All Latin names have a meaning, except for a few we have taken over from the ancient Greeks and Romans, but the most important thing about them is that they can be understood all over the world. Local names vary a great deal from one place to another even in the same country and so are not safe ways of identifying species. Latin names, however, do not vary. *Bellis perennis* means the plant we call the common Daisy, and it means the same plant in France, Turkey, China, or Australia.

CLASSIFYING PLANTS

There are untold millions of plants in the world — a thousand or so may be in one small back garden — and in order to study such large numbers they must be grouped together in some way. Plants with certain features in common are placed in the same group. The more characters they share, the closer is their relationship with each other. The most important features used in grouping — or classifying — plants are the number, shape, and arrangement of the parts of the flower and the type of fruit. The leaf shape and general appearance of plants are sometimes helpful, but these features tend to vary more from one plant to another. The size of a plant also varies a great deal according to the kind of place it is growing in and in a dry or wet season. The plant kingdom is divided into several major groups (1) the fungi: moulds, mushrooms, and toadstools; (2) the algae: seaweeds; (3) the mosses and liverworts; (4) the ferns and horsetails; (5) the conifers and flowering plants.

Within these major groups those plants with several characters in common are placed together in groups called families. For instance, the Field Cabbage (p. 11), Shepherd's Purse (p. 71), Charlock (p. 9), and Lady's Smock (p. 113) all have flowers with 4 petals and 4 long and 2 short stamens, so they are placed together with all other plants having these characters in the *Cruciferae*, or Cabbage family (p. 209). All plants with square stems, paired leaves, two-lipped, hooded flowers, and fruit consisting of 4 tiny nuts enclosed in the calyx are grouped together in the Deadnettle family, the *Labiatae* (p. 216).

Within each family the plants are divided into smaller groups. Each of these groups is called a genus, and the plants in each genus resemble each other more closely than they resemble the plants in another genus. For example, within the Deadnettle family all the plants in the Hemp-nettle genus (*Galeopsis*, p. 149) have two little knobs on the lower lip of the flower, and all the plants in the Calamint genus (*Calamintha*, p. 145) have stalked groups of flowers borne in the axils of the leaves. None of the other genera in the *Labiatae* family have these characters.

Within each genus the plants are divided into species. The plants in each species are so much alike that for practical purposes they cannot be distinguished. For example, among the Speedwells only the Germander Speedwell (p. 175) has two lines of hairs on the stem. All Germander Speedwells not only have this character but are very alike in all other respects.

Plants normally breed only with other plants in the same species. Sometimes, however, they may accidentally be pollinated by pollen from a different species. Often when this happens no seed is produced, but sometimes seeds will ripen and grow, and the resulting plant, called a hybrid, will have some characters of each parent. Often hybrid plants cannot produce seed and so do not reproduce themselves. A well-known example of a hybrid which does produce seed is the False Oxlip (p. 27), whose parents are the Cowslip and Primrose.

In naming plants the name of the genus comes first: thus all Speedwells are called *Veronica*, and the Germander Speedwell is called *Veronica chamaedrys*. The second name, the specific name, is given to the members of one species only in each genus; but the same specific name is often repeated for species in different genera (*see* NAMING PLANTS p. 206). A generic name is never repeated and so, together with the specific name, serves to distinguish a plant precisely.

FLOWER FAMILIES

There are altogether over 200 families of flowering plants, and about 120 are represented in Britain, though some with only one or with very rare species. We give here the main characteristics of about 45 of the biggest families, arranged in the probable order of evolution, the most primitive at the beginning and the more highly evolved at the end. The latter, such as *Compositae* and *Orchidaceae*, are the most easy to recognize because their features do not vary. The more primitive, such as *Ranunculaceae* or *Rosaceae*, vary not only from one genus or species to another but even from specimen to specimen. So they can be defined only in very general terms.

The most constant feature of most families is the form of the pistil: the number of compartments in the ovary, the number and arrangement of the seeds, and the position of the ovary above or below the sepals. A fruit with 4 nutlets, for instance, is either a Borage or a Labiate: one with two compartments and many seeds on the central axis belongs to either the *Scrophulariaceae* or the *Solanaceae*.

Other important points are the number and relationship of stamens, petals, and sepals. The families from *Ranunculaceae* to *Salicaceae*, for example, have petals free from one another. From *Ericaceae* to *Compositae* they are joined to form a tube. From *Liliaceae* to *Orchidaceae* the flower parts are in threes.

RANUNCULACEAE — Buttercup Family

All the plants in this family have numerous stamens and most have numerous carpels which are not joined together. Except for Baneberry (p. 67), the fruits are either single-seeded achenes (Buttercup, p. 3) or many-seeded follicles (Marsh Marigold, p. 5). The number of sepals and petals is not constant. The sepals are often coloured like the petals, as in the Wood Anemone (p. 67). In others the petals are modified to form nectaries, as in the Green and Stinking Hellebores (p. 51). Most *Ranunculaceae* flowers are symmetrical, but the garden flowers, Monkshood and Delphinium, are well known exceptions. Many species have a bitter taste and are poisonous. See pp. 3, 5.

Marsh Marigold (*Caltha palustris*)
Globe Flower (*Trollius europaeus*)
Green Hellebore (*Helleborus viridis*)
Baneberry (*Actaea spicata*)
Wood Anemone (*Anemone nemorosa*)
Pasque Flower (*A. pulsatilla*)
Traveller's Joy (*Clematis vitalba*)
Meadow Buttercup (*Ranunculus acris*)
Bulbous Buttercup (*R. bulbosus*)

Small-flowered Buttercup (*R. parviflorus*)
Goldilocks (*R. auricomus*)
Lesser Spearwort (*R. flammula*)
Celery-leaved Crowfoot (*R. sceleratus*)
Water Crowfoot (*R. aquatilis*)
Lesser Celandine (*R. ficaria*)
Pheasant's Eye (*Adonis annua*)
Mouse-tail (*Myosurus minimus*)
Common Meadow Rue (*Thalictrum flavum*)

PAPAVERACEAE — Poppy Family

All plants in this family contain a thick milky or coloured juice, which oozes out when any part of the plant is cut. The symmetrical flowers have 2 sepals which fall off as the flowers open, 4 petals, and numerous stamens. The fruit is a many-seeded capsule (p. 1, Fig. 3).

Field Poppy (*Papaver rhoeas*)
Welsh Poppy (*Meconopsis cambrica*)

Yellow Horned Poppy (*Glaucium flavum*)
Greater Celandine (*Chelidonium majus*)

CRUCIFERAE — Cabbage Family

This is an easy family to recognize as the 4 petals spread out like a Maltese cross. There are 4 sepals and 6 stamens, of which the outer 2 are shorter than the inner 4. The fruit is a specialized capsule, called a silicula if it is short and broad and a siliqua if it is long and slender. Many species are annuals and these include many familiar weeds and also plants grown as food crops. A few species such as Wall Rocket are slightly woody. See pp. 9, 11, 69, and 71.

Black Mustard (*Brassica nigra*)
Field Cabbage (*B. campestris*)
Charlock (*Sinapis arvensis*)
White Mustard (*S. alba*)
Wall Rocket (*Diplotaxis tenuifolia*)
Hedge Mustard (*Sisymbrium officinale*)
Treacle Mustard (*Erysimum cheiranthoides*)
Woad (*Isatis tinctoria*)
Common Winter Cress (*Barbarea vulgaris*)
Marsh Yellow-cress (*Rorippa islandica*)
Flixweed (*Descurainia sophia*)
Watercress (*Nasturtium officinale*)
Jack-by-the-hedge (*Alliaria petiolata*)

Hairy Bitter-cress (*Cardamine hirsuta*)
Lady's Smock (*C. pratensis*)
Wild Radish (*Raphanus raphanistrum*)
Sea Rocket (*Cakile maritima*)
Scurvy-grass (*Cochlearia officinalis*)
Field Penny-cress (*Thlaspi arvense*)
Hairy Rock-cress (*Arabis hirsuta*)
Shepherd's Purse (*Capsella bursa-pastoris*)
Thale Cress (*Arabidopsis thaliana*)
Pepperwort (*Lepidium campestre*)
Wart-cress (*Coronopus squamatus*)
Whitlow Grass (*Erophila verna*)

VIOLACEAE — Violet Family

The most characteristic feature of this family is the ovary which has a single cavity with three rows of ovules attached inside. The British species, all members of the violet genus, have a large lower petal with a pouch or spur projecting behind the sepals, and the two lower stamens are also spurred. The fruit, a capsule, splits on ripening into 3 boat-shaped segments, and the seeds are forced out. The leaves are simple and usually have stipules. See p. 165.

Common Dog Violet (*Viola riviniana*)
Wild Pansy (*V. tricolor*)
Sweet Violet (*V. odorata*)
Heath Dog Violet (*V. canina*)

Hairy Violet (*V. hirta*)
Marsh Violet (*V. palustris*)
Wood Dog Violet (*V. reichenbachiana*)

POLYGALACEAE — Milkwort Family

This family has unstalked leaves and irregular blue, pink, or white flowers. There are 5 sepals, the 2 inner being large and petal-like, and 3 to 5 smaller petals. The 8 stamens are in 2 bundles of 4, and there is 1 style. The ovary is 2-celled, and the fruit is a capsule.

Common Milkwort (*Polygala vulgaris*)
Heath Milkwort (*P. serpyllifolia*)

Chalk Milkwort (*P. calcarea*)

HYPERICACEAE — St. John's Wort Family

This family has untoothed opposite leaves, which are often marked with tiny transparent dots and never have stipules. The flowers have 5 sepals and petals and very numerous stamens grouped in 3 or 5 bundles. The fruit is a capsule. The family includes shrubs and non-woody perennial species.

Common St. John's Wort (*Hypericum perforatum*)
Slender St. John's Wort (*H. pulchrum*)

Trailing St. John's Wort (*H. humifusum*)

CARYOPHYLLACEAE — Pink Family

An easy family to recognize since the leaves are almost always paired and stalkless, and the stems are swollen at the nodes. There are usually 5 sepals, 5 petals, and 10 stamens, though some species, for example, the Pearlworts (p. 77), have only 4 petals or none at all. The 5 petals are sometimes notched and sometimes so deeply divided that at first glance it looks as though there are 10. There are two types of flower: one type is open and shallow with free sepals, as in the Chickweeds and Stitchworts; the other type is tubular, with joined sepals, as in the Campions and Pinks. The fruit is a capsule with the seeds attached to a central column. Most species are non-woody. See pp. 73, 75, 77, 107.

White Campion (*Melandrium album*)
Red Campion (*M. rubrum*)
Bladder Campion (*Silene cucubalus*)
Sea Campion (*S. maritima*)
Nottingham Catchfly (*S. nutans*)
Red Catchfly (*Viscaria vulgaris*)
Ragged Robin (*Lychnis flos-cuculi*)
Soapwort (*Saponaria officinalis*)
Corn Cockle (*Agrostemma githago*)
Chickweed (*Stellaria media*)
Greater Stitchwort (*S. holostea*)
Bog Stitchwort (*S. alsine*)
Sticky Mouse-ear Chickweed (*Cerastium glomeratum*)

Dark-green Mouse-ear Chickweed (*C. tetrandrum*)
Common Mouse-ear Chickweed (*C. vulgatum*)
Upright Chickweed (*Moenchia erecta*)
Knotted Pearlwort (*Sagina nodosa*)
Procumbent Pearlwort (*S. procumbens*)
Vernal Sandwort (*Minuartia verna*)
Sea Sandwort (*Honkenya peploides*)
Three-nerved Sandwort (*Moehringia trinervia*)
Thyme-leaved Sandwort (*Arenaria serpyllifolia*)
Corn Spurrey (*Spergula arvensis*)
Sea Spurrey (*Spergularia marginata*)
Water Chickweed (*Myosoton aquaticum*)

PORTULACACEAE — Purslane Family

The species in this small family can easily be mistaken for Chickweeds (p. 75), and most features of the flower, such as 5 regular petals and an ovary with a central column, are shared with the Caryophyllaceae. They can be distinguished, however, by the 2 sepals only and by the fewer stamens — 5 in Pink Purslane (p. 113) and 3 in Blinks (p. 85).

Blinks (*Montia verna*)

Pink Purslane (*Claytonia alsinoides*)

CHENOPODIACEAE — Goosefoot Family

The plants of this family are often fairly succulent or mealy, with alternate leaves and crowded insignificant greenish flowers. These tiny flowers have sepals only. Sometimes, as in the Goosefoots (p. 55) and Beet (p. 57), the male and female parts are in the same flower, though often they are in separate flowers, as in Orache (p. 55) and Sea Purslane (p. 57). The small green perianth of 3 to 5 sepals persists, enclosing the achene and frequently becoming fused to it. See pp. 55, 57.

Red Goosefoot (*Chenopodium rubrum*)
White Goosefoot (*C. album*)
Stinking Goosefoot (*C. vulvaria*)
Good King Henry (*C. bonus-henricus*)
Common Orache (*Atriplex patula*)
Grass-leaved Orache (*A. littoralis*)

Wild Beet (*Beta vulgaris*)
Sea Purslane (*Halimione portulacoides*)
Seablite (*Suaeda maritima*)
Saltwort (*Salsola kali*)
Glasswort (*Salicornia stricta*)

MALVACEAE — Mallow Family

A family easily recognized by the characteristic tuft of stamens projecting from the centre of the flower, and consisting of a tube made up of joined filaments, which frays at the end into many fine branches, each bearing a single anther cell. The ovary in British species is a flattened disc which, when ripe, splits vertically from the centre into many flat crescent-shaped slices, each containing a single seed (p. 131). This is very clear in the large fruits of the garden Hollyhock. There are 5 sepals, usually with an additional outer cup of 3 or more, and 5 free petals, often notched.

Common Mallow (*Malva sylvestris*)

Musk Mallow (*M. moschata*)

GERANIACEAE — Cranesbill Family

The most conspicuous feature of this family is the fruit — a capsule with a long beak, formed by the style and the seeds at its base. There are 5 sepals and petals and 10 stamens, with the outer 5 opposite the petals and the inner 5 opposite the sepals. Most species are non-woody and have deeply lobed leaves with stipules at their base. See p. 129.

Meadow Cranesbill (*Geranium pratense*)
Bloody Cranesbill (*G. sanguineum*)
Cut-leaved Cranesbill (*G. dissectum*)
Dove's-foot Cranesbill (*G. molle*)

Shining Cranesbill (*G. lucidum*)
Herb Robert (*G. robertianum*)
Common Storksbill (*Erodium cicutarium*)

PAPILIONACEAE — Pea Family

The most distinctive feature of this family is the flower shape. The sepals are united into a tube. There are 5 petals; the broad upper one is called the standard, the 2 side ones the wings, and the 2 lower ones the keel. The keel petals, which are often joined at their lower edges, enclose the 10 stamens. The stamens are fused into a tube, either 9 of them or all 10, and this tube encloses the carpel which later forms the pod. The leaves have stipules and are usually compound, sometimes forming tendrils. The family includes a wide variety of plants from trees to tiny Vetches and Clovers. See pp. 19, 21, 23, 115, 133, 135.

Dyer's Greenweed (*Genista tinctoria*)
Broom (*Sarothamnus scoparius*)
Gorse (*Ulex europaeus*)
Dwarf Furze (*U. minor*)
Common Birdsfoot (*Ornithopus perpusillus*)
Birdsfoot-trefoil (*Lotus corniculatus*)
Least Birdsfoot-trefoil (*L. angustissimus*)
Hop Trefoil (*Trifolium campestre*)
Red Clover (*T. pratense*)
Zigzag Clover (*T. medium*)
Sea Clover (*T. squamosum*)
Hare's-foot (*T. arvense*)
Strawberry Clover (*T. fragiferum*)
Alsike Clover (*T. hybridum*)
White Clover (*T. repens*)
Black Medick (*Medicago lupulina*)
Spotted Medick (*M. arabica*)

Hairy Medick (*M. hispida*)
Lucerne (*M. sativa*)
Yellow Vetchling (*Lathyrus aphaca*)
Grass Vetchling (*L. nissolia*)
Everlasting Pea (*L. sylvestris*)
Bitter Vetch (*L. montanus*)
Sea Pea (*L. maritimus*)
Meadow Vetchling (*L. pratensis*)
Kidney-vetch (*Anthyllis vulneraria*)
Horse-shoe Vetch (*Hippocrepis comosa*)
Milk Vetch (*Astragalus glycyphyllos*)
Melilot (*Melilotus officinalis*)
Restharrow (*Ononis repens*)
Hairy Tare (*Vicia hirsuta*)
Tufted Vetch (*V. cracca*)
Bush Vetch (*V. sepium*)
Common Vetch (*V. sativa*)

ROSACEAE — Rose Family

This large and varied family has stipules on the leaves and regular or symmetrical flowers. Indeed, these are the only constant features of the family. The petals and sepals are usually 5, and the stamens, if not too numerous to count, are in whorls of 5 or 10: in Cinquefoil (p. 17) there are 20 or 30. The receptacle varies greatly; it may be cone-shaped as in Avens (p. 17), saucer-shaped as in Meadow-sweet (p. 81), or urn-shaped — either dry as in Agrimony (p. 17) or fleshy as in the hip, haw, or apple (p. 181). The carpels are usually numerous and free, ripening to give a cluster of drupes, achenes, or follicles. More rarely there is only 1 carpel, which either dries to form a single achene (Burnet, p. 157) or becomes fleshy and contains a stone (Blackthorn, p. 181). See also pp. 17, 79, 81, 117, 195.

Herb Bennet (*Geum urbanum*)
Water Avens (*G. rivale*)
Mountain Avens (*Dryas octopetala*)
Silverweed (*Potentilla anserina*)
Shrubby Cinquefoil (*P. fruticosa*)
Creeping Cinquefoil (*P. reptans*)
Marsh Cinquefoil (*P. palustris*)
Barren Strawberry (*P. sterilis*)
Tormentil (*P. erecta*)
Agrimony (*Agrimonia eupatoria*)
Wild Strawberry (*Fragaria vesca*)
Lady's Mantle (*Alchemilla vulgaris*)
Field Rose (*Rosa arvensis*)
Burnet Rose (*R. spinosissima*)
Dog Rose (*R. canina* agg.)
Sweet-briar (*R. rubiginosa* agg.)

Cloudberry (*Rubus chamaemorus*)
Blackberry (*R. fruticosus*)
Raspberry (*R. idaeus*)
Dewberry (*R. caesius*)
Meadow-sweet (*Filipendula ulmaria*)
Great Burnet (*Sanguisorba officinalis*)
Salad Burnet (*Poterium sanguisorba*)
Blackthorn (*Prunus spinosa*)
Cherry-plum (*P. cerasifera*)
Wild Cherry (*P. avium*)
Bird Cherry (*P. padus*)
Hawthorn (*Crataegus monogyna*)
Crab Apple (*Malus sylvestris*)
Wild Service Tree (*Sorbus torminalis*)
White Beam (*S. aria* agg.)
Rowan (*S. aucuparia*)

CRASSULACEAE — Stonecrop Family

This family consists mainly of small plants with succulent leaves, and this feature, together with the free (or almost free) carpels equal in number to the petals, makes the family easy to recognize. The petals are frequently 5, though the Stonecrops (pp. 15, 83, 113) may have as many as 10. The stamens either correspond in number with the petals, alternating with them, or they are twice as numerous, as in the Stonecrops.

Roseroot (*Sedum rosea*)
Yellow Stonecrop (*S. acre*)
English Stonecrop (*S. anglicum*)

Hairy Stonecrop (*S. villosum*)
Wall Pennywort (*Umbilicus rupestris*)

SAXIFRAGACEAE — Saxifrage Family

This family consists of small, neat, non-woody plants, often with alternate leaves. The receptacle is saucer-shaped, and the 2 carpels, joined at the base and free above, distinguish it fairly easily from the Crassulaceae. Petals are usually 5, though in Golden Saxifrage (p. 15) they are absent and there are only 4 sepals. In the British species there are twice as many stamens as petals.

Mossy Saxifrage (*Saxifraga hypnoides*)
Meadow Saxifrage (*S. granulata*)

Golden Saxifrage (*Chrysosplenium oppositifolium*)

LYTHRACEAE — Loosestrife Family

The family is distinguished by an unusual combination of joined sepals, and free petals. The petals, 4-6 in Purple Loosestrife (p. 123) and 6 in Water Purslane (p. 113), are joined to the inside of the calyx tube, and are crumpled in the bud. The stamens, often twice as many as the petals, are inserted lower down on the calyx in one or two whorls. The fruit in British species is a capsule, with 2 cavities.

Purple Loosestrife (*Lythrum salicaria*)

Water Purslane (*Peplis portula*)

ONAGRACEAE — Willow-herb Family

These non-woody plants are often found in wet places. The flowers have 4 petals and sepals, and can be distinguished from the few other families with 4 petals (notably *Cruciferae* and *Rubiaceae*) by their 8 stamens (rarely 2 or 4), and by the position below the rest of the flower of the ovary with its 2 or 4 compartments and rows of small seeds. Enchanter's Nightshade (p. 95) is rather exceptional in having only 2 sepals, petals, and stamens and an ovary with 2 compartments, each containing 1 or 2 seeds. See also p. 111.

Broad-leaved Willow-herb (*Epilobium montanum*)
Great Hairy Willow-herb (*E. hirsutum*)
Marsh Willow-herb (*E. palustre*)

New Zealand Willow-herb (*E. pedunculare*)
Rosebay Willow-herb (*Chamaenerion angustifolium*)
Enchanter's Nightshade (*Circaea lutetiana*)

UMBELLIFERAE — Parsley Family

The arrangement of the tiny flowers in groups called umbels gives this family its name. An umbel is a group of flowers with stalks all growing from the top of the stem, so that the whole group looks like an umbrella (*see* p. 46, fig. 1). The flowers have 5 sepals, petals, and stamens. The leaves are usually divided and often fern-like, and the stems are often furrowed. Most plants have a tap-root, which, as in the Carrot and Parsnip, may be swollen. See pp. 47, 87, 89, 91.

Sanicle (*Sanicula europaea*)
Wild Angelica (*Angelica sylvestris*)
Celery (*Apium graveolens*)
Fool's Watercress (*A. nodiflorum*)
Fine-leaved Marshwort (*A. inundatum*)
Alexanders (*Smyrnium olusatrum*)
Sea Holly (*Eryngium maritimum*)
Rock Samphire (*Crithmum maritimum*)
Fennel (*Foeniculum vulgare*)
Rough Chervil (*Chaerophyllum temulum*)
Cow Parsley (*Anthriscus sylvestris*)
Shepherd's Needle (*Scandix pecten-veneris*)
Upright Hedge-parsley (*Torilis japonica*)
Knotted Hedge-parsley (*T. nodosa*)

Hemlock (*Conium maculatum*)
Pignut (*Conopodium majus*)
Burnet Saxifrage (*Pimpinella saxifraga)*
Ground Elder (*Aegopodium podagraria*)
Narrow-leaved Water-parsnip (*Berula erecta*)
Water Dropwort (*Oenanthe fistulosa*)
Hemlock Water Dropwort (*Oe. crocata*)
Parsley Water Dropwort (*Oe. lachenalii*)
Pepper Saxifrage (*Silaum silaus*)
Wild Parsnip (*Pastinaca sativa*)
Hogweed (*Heracleum sphondylium*)
Wild Carrot (*Daucus carota*)
Fool's Parsley (*Aethusa cynapium*)
Marsh Pennywort (*Hydrocotyle vulgaris*)

POLYGONACEAE — Dock Family

This family at first sight looks rather like the Goosefoots (p. 55), owing to the complicated spikes of small greenish flowers. It can be distinguished from the Chenopodiaceae, however, by the stipules on the leaves which form a sheath round the stem. There is no real distinction between petals and sepals, though the perianth may be coloured as in the Bistorts and Persicarias (p. 127), and in the Docks (p. 59) large and small green sepals alternate. The perianth persists as the one-seeded fruit hardens. If there are 3 main sepals the fruit becomes triangular as in the Docks, and if only 2 it becomes flattened and lens-shaped as in Mountain Sorrel (p. 59). There are 6-9 stamens.

Knotgrass (*Polygonum aviculare*)
Bistort (*P. bistorta*)
Amphibious Bistort (*P. amphibium*)
Spotted Persicaria (*P. persicaria*)
Water Pepper (*P. hydropiper*)
Black Bindweed (*P. convolvulus*)

Mountain Sorrel (*Oxyria digyna*)
Common Sorrel (*Rumex acetosa*)
Sheep's Sorrel (*R. acetosella*)
Curled Dock (*R. crispus*)
Broad Dock (*R. obtusifolius*)
Sharp Dock (*R. conglomeratus*)

URTICACEAE — Nettle Family

This, like the Docks (p. 59) and Goosefoots (p. 55), is a family of plants with insignificant green flowers. Many of the Nettles have stinging hairs on leaves and stems. Except in Pellitory-of-the-wall (p. 65) male and female flowers are separate. In the male flowers the stamens, which explode to release the pollen, are opposite the 4 sepals, and there is an undeveloped pistil. In the female flowers, the stamens are rudimentary, and the pistil develops to give a one-celled dry or fleshy fruit, containing a single seed.

Stinging Nettle (*Urtica dioica*)

Pellitory-of-the-wall (*Parietaria diffusa*)

EUPHORBIACEAE — Spurge Family

This flourishing family of largely tropical plants has few British representatives. The small, insignificant flowers have no petals, sometimes no sepals, and male and female parts in different flowers. The ovary has 3 compartments, each containing one or 2 seeds, and splits into 3 segments when ripe, leaving a central rod. Dog's Mercury (p. 63) has 3 sepals enclosing either stamens or ovary, whereas the Spurges (p. 63) have instead of sepals a special green cup or involucre, enclosing a group of male flowers and a single female flower. Most Spurges have a milky juice, called latex, which in some tropical species is the source of rubber.

Dog's Mercury (*Mercurialis perennis*)   Wood Spurge (*E. amygdaloides*)
Sun Spurge (*Euphorbia helioscopia*)

FAGACEAE — Beech Family

A family of trees or shrubs with simple, alternate leaves and separate, much reduced male and female flowers. A persistent scaly cup or sheath surrounds the female flowers: in the Beech (p. 201) and Sweet Chestnut, where the female flowers are in clusters of 2 or 3, this forms a spiky sheath surrounding the nuts; in the Oak, where the flower is single, this is the acorn cup (p. 201). The male flowers form catkins or tassel-like heads, each flower consisting of a simple perianth of 4-6 lobes, enclosing at least twice as many stamens.

Beech (*Fagus sylvatica*)   Oak (*Quercus robur*)

SALICACEAE — Willow Family

This family of trees or shrubs has stipules at the base of the leaf stalks, and tiny reduced flowers packed together in catkins. Male and female plants are separate. The individual flowers in the conspicuous male catkins consist of bracts containing stamens. Poplars have more stamens than Willows, where there are generally only 2. Each flower in the female catkins consists of a bract containing an ovary with 1 compartment, and seeds arranged in 2 or 4 rows. The fruit is a capsule and opens by two valves to expose a fluff of silky hairs surrounding the seeds.

Poplars are wind pollinated, but Willows, where both male and female flowers contain 1 or 2 small nectaries, are insect-pollinated. The Willows form hybrids or crosses very freely, and intermediate forms between two species are often found. See also p. 178.

White Poplar (*Populus alba*)   Great Sallow (*S. caprea*)
Aspen (*P. tremula*)   Common Osier (*S. viminalis*)
Black Poplar (*P. nigra*)   Crack Willow (*S. fragilis*)
Purple Willow (*Salix purpurea*)   Creeping Willow (*S. repens*)

ERICACEAE — Heather Family

The plants in this family are almost all shrubs with thick, leathery, and often very small leaves. Except for some Rhododendrons, the flowers are rather bell-like, and waxy. In all *Ericaceae* the pollen is shed through pores at the top of the anthers instead of through slits, as in other families. All the British species, except the very rare Strawberry Tree, are only found on lime-free soils. See p. 119.

Cross-leaved Heath (*Erica tetralix*)   Bearberry (*Arctostaphylos uva-ursi*)
Dorset Heath (*E. ciliaris*)   Cowberry (*Vaccinium vitis-idaea*)
Bell Heather (*E. cinerea*)   Bilberry (*V. myrtillus*)
Ling (*Calluna vulgaris*)   Cranberry (*Oxycoccus palustris*)
Bog Rosemary (*Andromeda polifolia*)

PLUMBAGINACEAE — Sea Lavender Family

The British members of this family are most easily recognized by the papery, coloured, pleated, tubular calyx which persists round the fruit and forms the "everlasting flowers" of the cultivated Statice. The flowers have 5 equal petals commonly joined to form a tube, though in Thrift (p. 127) they are free almost to the base. The 5 stamens are joined to the corolla opposite the lobes. The ovary has a single cavity containing one seed, and bearing 5 more or less separate stigmas. These are plants of the sea-shore.

Sea Lavender (*Limonium vulgare*)   Thrift (*Armeria maritima*)

PRIMULACEAE — Primrose Family

Non-woody plants with leaves without stipules, and symmetrical flowers. The 5 petals are joined at the base, sometimes as in the Primrose and Cowslip into a long tube. The calyx is also often tubular. The 5 stamens are always opposite the petals. The fruit is a capsule. See p. 27.

Primrose (*Primula vulgaris*)
Cowslip (*P. veris*)
Bird's Eye Primrose (*P. farinosa*)
Oxlip (*P. elatior*)
False Oxlip (*P. veris x vulgaris*)
Yellow Loosestrife (*Lysimachia vulgaris*)
Creeping Jenny (*L. nummularia*)

Yellow Pimpernel (*L. nemorum*)
Water Violet (*Hottonia palustris*)
Bog Pimpernel (*Anagallis tenella*)
Scarlet Pimpernel (*A. arvensis*)
Chaffweed (*Centunculus minimus*)
Sea Milkwort (*Glaux maritima*)
Brookweed (*Samolus valerandi*)

OLEACEAE — Ash Family

This family of trees or shrubs includes the well-known Olive of warm climates. The flowers are regular, usually with either a 4-lobed calyx and corolla or none, 2 stamens, 1 style, and a 2-celled ovary. There are two genera: (1) with pinnate leaves and winged fruits (Ash, p. 197); (2) with simple leaves and black berry fruits (Privet, p. 183).

Ash (*Fraxinus excelsior*)

Privet (*Ligustrum vulgare*)

GENTIANACEAE — Gentian Family

This is a well-defined group of non-woody plants with opposite leaves. There are usually 5 regular petals, or sometimes 4 as in Field Gentian (p. 139), and these are joined to form a tube to which the 5 (4) stamens are attached. The stamens are joined to the corolla tube between the lobes. The fruit is most characteristic, having one cavity with two rows of seeds inside attached to the lining.

Field Gentian (*Gentianella campestris*)
Felwort (*G. amarella*)
Spring Gentian (*Gentiana verna*)
Marsh Gentian (*G. pneumonanthe*)

Common Centaury (*Centaurium minus*)
Dumpy Centaury (*C. capitatum*)
Yellow-wort (*Blackstonia perfoliata*)

BORAGINACEAE — Forget-me-not Family

Almost all the plants in this family are covered with short, stiff hairs so that they feel rough or even bristly. The flowers are often blue and have 5 sepals, petals, and stamens. The main flower stalk is tightly curled in bud, uncurling as the flowers open, so that the newly-opened flowers always face upward or outward. The fruit consists of 4 tiny nuts enclosed by the calyx. See pp. 171, 173.

Water Forget-me-not (*Myosotis palustris*)
Wood Forget-me-not (*M. sylvatica*)
Common Forget-me-not (*M. arvensis*)
Changing Forget-me-not (*M. discolor*)
Early Forget-me-not (*M. hispida*)
Borage (*Borago officinalis*)
Alkanet (*Pentaglottis sempervirens*)

Bugloss (*Lycopsis arvensis*)
Viper's Bugloss (*Echium vulgare*)
Blue Gromwell (*Lithospermum purpurocaeruleum*)
Sea Lungwort (*Mertensia maritima*)
Gromwell (*Lithospermum officinale*)
Comfrey (*Symphytum officinale*)
Hound's-tongue (*Cynoglossum officinale*)

CONVOLVULACEAE — Bindweed Family

These are characteristically climbing plants. The flower has a spreading trumpet-shaped corolla, which may be divided shallowly into 5 lobes, with the 5 stamens arising between the lobes. The fruit is a capsule and splits into 2 or 4 valves.

Larger Bindweed (*Calystegia sepium*)
Small Bindweed (*Convolvulus arvensis*)

Common Dodder (*Cuscuta epithymum*)

SOLANACEAE — Nightshade Family

A family closely related to the Scrophulariaceae and having the same joined calyx and corolla and an ovary with 2 cells and many ovules. The corolla of the Solanaceae, however, has equal lobes, giving a perfect regularity to the flower. There are usually 5 lobes with the 5 stamens inserted between them. The fruit may be a berry, as in Woody Nightshade (p. 131), or a capsule, as in Henbane (p. 29). Many plants of this family are highly poisonous, notably Deadly Nightshade (p. 131).

Duke of Argyll's Tea-plant (*Lycium halimifolium*)
Deadly Nightshade (*Atropa belladonna*)
Woody Nightshade (*Solanum dulcamara*)

Black Nightshade (*S. nigrum*)
Henbane (*Hyoscyamus niger*)

SCROPHULARIACEAE — Snapdragon Family

This family has some characters resembling the Labiatae but can always be distinguished by the fruit, which is a capsule. The corolla is of several types: shallow with spreading, slightly unequal lobes, either 5 lobes and stamens (Mulleins, p. 29) or 2 stamens and 4 lobes (Speedwells, p. 175); hooded like the Labiatae (Rattles and Lousewort, (p. 141); and unhooded, either with a closed mouth (Toadflax, p. 141) or an open mouth (Foxglove, p. 123). See pp. 25, 141, 175.

Great Mullein (*Verbascum thapsus*)
Common Cow-wheat (*Melampyrum pratense*)
Yellow Rattle (*Rhinanthus minor*)
Common Toadflax (*Linaria vulgaris*)
Round-leaved Fluellen (*Kickxia spuria*)
Monkey Flower (*Mimulus guttatus*)
Foxglove (*Digitalis purpurea*)
Eyebright (*Euphrasia officinalis* agg.)
Small Toadflax (*Chaenorrhinum minus*)
Ivy-leaved Toadflax (*Cymbalaria muralis*)
Figwort (*Scrophularia nodosa*)

Red Bartsia (*Odontites verna*)
Red-rattle (*Pedicularis palustris*)
Lousewort (*P. sylvatica*)
Brooklime (*Veronica beccabunga*)
Water Speedwell (*V. anagallis-aquatica*)
Common Speedwell (*V. officinalis*)
Germander Speedwell (*V. chamaedrys*)
Thyme-leaved Speedwell (*V. serpyllifolia*)
Wall Speedwell (*V. arvensis*)
Large Field Speedwell (*V. persica*)
Marsh Speedwell (*V. scutellata*)

OROBANCHACEAE — Broomrape Family

The plants in this family are parasites on the roots of other plants. There are no green leaves, these having been reduced to scales on the flower stalk. The flowers have 2-4-lobed calyxes, irregular, more or less 2-lipped corollas, 4 stamens, and 1 style. The fruit is a 2-valved capsule.

Toothwort (*Lathraea squamaria*)

Lesser Broomrape (*Orobanche minor*)

LABIATAE — Deadnettle Family

Plants in this family all have square stems and leaves in pairs. The 5 sepals are joined and persist to enclose the fruit, which consists of 4 little nuts. The 5 petals are joined at the base to form a tube. At the top they divide into 2 lobes, the upper lobe, or hood, being formed by 2 petals and the lower lip by 3. The 4 stamens are protected by the hood. Most members of the family are non-woody, and many have a characteristic smell, either pleasant or otherwise. See pp. 97, 143, 145, 147, 149.

Penny-royal (*Mentha pulegium*)
Corn Mint (*M. arvensis*)
Water Mint (*M. aquatica*)
Horse-mint (*M. longifolia*)
Gipsywort (*Lycopus europaeus*)
Marjoram (*Origanum vulgare*)
Wild Thyme (*Thymus drucei*)
Common Calamint (*Calamintha ascendens*)
Basil-thyme (*Acinos arvensis*)
Wild Basil (*Clinopodium vulgare*)
Meadow Clary (*Salvia pratensis*)
Wild Clary (*S. horminoides*)
Bastard Balm (*Melittis melissophyllum*)
Self-heal (*Prunella vulgaris*)
Cut-leaved Self-heal (*P. laciniata*)
Betony (*Stachys officinalis*)

Marsh Woundwort (*S. palustris*)
Hedge Woundwort (*S. sylvatica*)
Black Horehound (*Ballota nigra*)
Yellow Archangel (*Galeobdolon luteum*)
Henbit Deadnettle (*Lamium amplexicaule*)
Red Deadnettle (*L. purpureum*)
White Deadnettle (*L. album*)
Red Hemp-nettle (*Galeopsis angustifolia*)
Common Hemp-nettle (*G. tetrahit*)
Ground Ivy (*Glechoma hederacea*)
White Horehound (*Marrubium vulgare*)
Common Skullcap (*Scutellaria galericulata*)
Lesser Skullcap (*S. minor*)
Wood Sage (*Teucrium scorodonia*)
Ground-pine (*Ajuga chamaepitys*)
Bugle (*A. reptans*)

PLANTAGINACEAE — Plantain Family

These herbaceous plants have leaves almost invariably growing from the base of the stem, and leafless flowering stems with a spike of closely packed flowers at the tip. The corolla has 4 small spreading papery lobes, and there are 4 persistent sepals united at the base and 4 long stamens. In all the Plantains the stamens are fixed to the corolla tube and the ovary has 2 to 4 compartments. The fruit is a capsule which splits round its circumference, the top half coming off as a cap. See p. 61.

Greater Plantain (*Plantago major*)
Hoary Plantain (*P. media*)

Buck's-horn Plantain (*P. coronopus*)
Ribwort Plantain (*P. lanceolata*)

CAMPANULACEAE — Bellflower Family

The feature which gives the name to this family is the bell-shaped corolla, familiar in the Harebell. In some species, however, such as Rampion or Sheep's-bit (p. 179), the flowers are gathered into a head rather like that of a Composite, but distinguished from the Compositae by the free anthers. The ovary, which is below the sepals (inferior), is 3-5 celled (2 celled in Sheep's-bit), with a corresponding number of stigmas. The fruit is always a capsule in which, in British species, pores or valves open to allow numerous small seeds to escape. The flowers of this family are most frequently blue. See p. 167.

Giant Bellflower (*Campanula latifolia*)
Nettle-leaved Bellflower (*C. trachelium*)
Creeping Bellflower (*C. rapunculoides*)
Clustered Bellflower (*C. glomerata*)

Spreading Bellflower (*G. patula*)
Harebell (*C. rotundifolia*)
Round-headed Rampion (*Phyteuma tenerum*)
Sheep's-bit (*Jasione montana*)

RUBIACEAE — Bedstraw Family

In all the British members of this family the leaves are arranged in whorls of at least 4 leaves: for example, 4 in Crosswort and 8-12 in Yellow Bedstraw (p. 31). Only 2 in each whorl are true leaves, with buds in the axils, the remainder usually representing enlarged stipules. The flower corolla is characteristically cruciform, with 4 stamens alternating with the 4 small spreading lobes of the corolla — though Wild Madder (p. 49) has 5 lobes and stamens. The ovary is below the sepals and, except for Wild Madder, has 2 cells. When ripe the fruit separates into 2 parts, each containing a single seed. In Wild Madder only one cell develops.

Yellow Bedstraw (*Galium verum*)
Heath Bedstraw (*G. hercynicum*)
Hedge Bedstraw (*G. mollugo*)
Crosswort (*G. cruciata*)
Goosegrass (*G. aparine*)

Woodruff (*Asperula odorata*)
Squinancy Wort (*A. cynanchica*)
Field Madder (*Sherardia arvensis*)
Wild Madder (*Rubia peregrina*)

CAPRIFOLIACEAE — Honeysuckle Family

A family of shrubby plants, with opposite leaves and a woody growth. The flower has 5 corolla lobes, usually equal and regular, but in Honeysuckle the corolla has two lips, the upper 4-lobed, and the lower with only 1. The ovary, which is below the sepals, has from 1 to 5 fertile carpels, giving a fruit which may be an achene, a stone with a fleshy cover (Wayfaring Tree, p. 193), or a berry (Honeysuckle, p. 15).

Honeysuckle (*Lonicera periclymenum*)
Elder (*Sambucus nigra*)

Wayfaring Tree (*Viburnum lantana*)
Guelder Rose (*V. opulus*)

VALERIANACEAE — Valerian Family

This is a small family of plants with opposite leaves and small flowers, often gathered into loose heads. The corolla may be slightly irregular (asymmetrical), sometimes having a small pouch, as in the Common Valerian, or a spur, as in the Red Valerian, at the base of the tube. Stamens are 3 (except in the Red Valerian which has 1) and joined to the base of the corolla tube. The ovary has 3 cells, but only one is fertile, and this contains a single seed. In both Common and Red Valerian (p. 123) a tuft or pappus of hairs develops on the fruit from the small rolled calyx at the top of the ovary.

Common Valerian (*Valeriana officinalis*)
Red Valerian (*Kentranthus ruber*)

Corn Salad (*Valerianella locusta*)

DIPSACACEAE — Scabious Family

This family, like the Compositae, has a large number of small flowers or florets collected together in a cup or sheath of bracts to form a 'head', often mistaken for a single flower. Each individual floret has, in addition to the reduced calyx at the top of the ovary, a distinctive extra calyx at the base which persists and encloses the fruit. In the Teasels (p. 157) this second calyx is conspicuous and spiny. The corolla tube is usually curved with 4 or 5 slightly unequal lobes. The 2 or 4 stamens are joined to the corolla, their free anthers distinguishing the Dipsacaceae from the Compositae. The fruit is dry and contains a single seed.

Teasel (*Dipsacus fullonum*)
Field Scabious (*Knautia arvensis*)

Small Scabious (*Scabiosa columbaria*)
Devil's-bit Scabious (*Succisa pratensis*)

COMPOSITAE — Daisy Family

This most abundant and successful of all flowering families can always be recognized by a combination of two features. The first is its composite flower-heads — a characteristic shared with the Dipsacaceae (p. 157) and some Campanulaceae (p. 179). The second, found only in the Compositae, is that the 5 anthers are joined side to side to form a cylinder, into which the pollen is shed, and through which the style grows. The petals of each floret are joined to form either a tube, or a strap. Sometimes both types of floret are found in a flower-head, as in the Daisy (p. 99); sometimes only strap-shaped florets, as in the Dandelion (p. 41); and sometimes only tubular ones, as in Tansy (p. 35). The fruit contains a single dry seed which is often carried on the wind by a parachute of hairs at the apex, representing a much reduced calyx. Many Compositae have a milky juice in the stems and leaves. See pp. 33, 35, 37, 39, 41, 43, 99, 151, 153, 155.

Nodding Bur-marigold (*Bidens cernuus*)
Common Cudweed (*Filago germanica*)
Slender Cudweed (*F. minima*)
Wayside Cudweed (*Gnaphalium uliginosum*)
Golden Rod (*Solidago virgaurea*)
Wormwood (*Artemisia absinthium*)
Mugwort (*A. vulgaris*)
Sea Wormwood (*A. maritima*)
Common Ragwort (*Senecio jacobaea*)
Marsh Ragwort (*S. aquaticus*)
Oxford Ragwort (*S. squalidus*)
Wood Groundsel (*S. sylvaticus*)
Common Groundsel (*S. vulgaris*)
Coltsfoot (*Tussilago farfara*)
Butterbur (*Petasites hybridus*)
Ploughman's Spikenard (*Inula conyza*)
Golden Samphire (*I. crithmoides*)
Common Fleabane (*Pulicaria dysenterica*)
Cat's-foot (*Antennaria dioica*)
Sea Aster (*Aster tripolium*)
Blue Fleabane (*Erigeron acris*)
Daisy (*Bellis perennis*)
Hemp Agrimony (*Eupatorium cannabinum*)
Corn Chamomile (*Anthemis arvensis*)
Scentless Mayweed (*Matricaria maritima*)
Rayless Mayweed (*M. matricarioides*)
Yarrow (*Achillea millefolium*)
Sneezewort (*A. ptarmica*)
Corn Marigold (*Chrysanthemum segetum*)
Ox-eye Daisy (*C. leucanthemum*)
Feverfew (*C. parthenium*)

Tansy (*Tanacetum vulgare*)
Carline Thistle (*Carlina vulgaris*)
Great Burdock (*Arctium lappa*)
Slender Thistle (*Carduus tenuiflorus*)
Musk Thistle (*C. nutans*)
Welted Thistle (*C. crispus*)
Woolly Thistle (*Cirsium eriophorum*)
Spear Thistle (*C. vulgare*)
Marsh Thistle (*C. palustre*)
Creeping Thistle (*C. arvense*)
Stemless Thistle (*C. acaule*)
Meadow Thistle (*C. dissectum*)
Scotch Thistle (*Onopordum acanthium*)
Greater Knapweed (*Centaurea scabiosa*)
Lesser Knapweed (*C. nigra*)
Cornflower (*C. cyanus*)
Saw-wort (*Serratula tinctoria*)
Chicory (*Cichorium intybus*)
Nipplewort (*Lapsana communis*)
Common Cat's Ear (*Hypochaeris radicata*)
Hawkbits (*Leontodon autumnalis, hispidus*)
Bristly Ox-tongue (*Picris echioides*)
Goat's-beard (*Tragopogon pratensis*)
Wall Lettuce (*Lactuca muralis*)
Corn Sowthistle (*Sonchus arvensis*)
Common Sowthistle (*S. oleraceus*)
Hawkweeds (*Hieracium*: groups *Aphyllopoda, Phyllopoda, Pilosella*)
Orange Hawkweed (*H. aurantiacum*)
Hawk's-beards (*Crepis taraxacifolia, capillaris*)
Dandelion (*Taraxacum officinale*)

HYDROCHARITACEAE — Frogbit Family

A very small group of aquatic plants, which are partly or wholly submerged, even pollination sometimes taking place under water. Several male flowers and one female flower are enclosed in a forked bract or spathe. There are 3 sepals and 3 petals, though the latter are sometimes small or even, as in Canadian Pondweed (p. 53), absent. Stamens are usually 3, or multiples of three: Water Soldier (p. 103) has 12. The ovary is below the sepals and has either a single cavity with 3 rows of seeds, as in Canadian Pondweed, or 3, 6, or 9 compartments, usually 6 as in Frogbit (p. 103). The small fruit ripens under water and does not open.

Frogbit (*Hydrocharis morsus-ranae*)
Water Soldier (*Stratiotes aloides*)

Canadian Pondweed (*Elodea canadensis*)

LILIACEAE — Lily Family

With few exceptions, the flowers in this family and of a number of related families have their parts in multiples of 3. Another feature is the parallel veins in the leaves. In the Liliaceae there are 6 petal-like parts of the perianth in 2 whorls of 3, with a stamen opposite each. The ovary within the perianth has 3 compartments, with numerous ovules on a central axis. The fruit is a capsule, or less commonly a berry.

Turk's Cap Lily (*Lilium martagon*)
Fritillary (*Fritillaria meleagris*)
Spring Squill (*Scilla verna*)
Bluebell (*Endymion nonscriptus*)
Bog Asphodel (*Narthecium ossifragum*)

Crow Garlic (*Allium vineale*)
Ramsons (*A. ursinum*)
Meadow Saffron (*Colchicum autumnale*)
Herb Paris (*Paris quadrifolia*)

AMARYLLIDACEAE — Daffodil Family

This family, with few British members, differs from Liliaceae in the papery sheath or spathe which encloses the flower bud, and in the position of the ovary below the perianth. In the Daffodil (p. 29), the perianth has a trumpet enclosing the stamens; in other species the inner 3 sections of the perianth are smaller than the outer 3. All members of this family develop bulbs from which the leaves grow directly. The flowering stems are leafless.

Daffodil (*Narcissus pseudonarcissus*)

IRIDACEAE — Iris Family

This family differs from Liliaceae in the position of the ovary below the perianth (inferior), and from both Liliaceae and Amaryllidaceae in having only 3 stamens. The 6 parts of the perianth are in two series of 3, both coloured like petals and joined at the base to form a short tube. The ovary, as in Liliaceae, has 3 compartments with ovules growing on the central axis. The style has 3 lobes which are sometimes large and petal-like. Storage organs, in the form of bulbs or swollen stems, are always formed. These may creep horizontally as in Yellow Flag (p. 29), or may be bulbous and upright, forming a corm, as in Autumn Crocus (p. 163).

Gladdon (*Iris foetidissima*)
Yellow Flag (*I. pseudacorus*)

Autumn Crocus (*Crocus nudiflorus*)

ORCHIDACEAE — Orchid Family

This large family of herbaceous plants has highly complex flowers, adapted to special methods of pollination. As in Liliaceae, the perianth is in 6 parts, usually all coloured. Of the inner 3, the much enlarged lower one, the 'labellum', serves as a landing stage for insects. In all but one rare genus there are 3 stigmas, 2 fertile and one sterile, and only one stamen. Both stamen and stigmas are attached to a fleshy central column. Except in the one genus, the minute pollen grains are bound together by elastic threads to form club-shaped masses or 'pollinia'. In some species, for example, the Early Purple Orchis (p. 159), the pollinia end in sticky discs; when an insect presses against these, the ripe pollinia are dragged out whole from the open anther cells. The ovary, which is below the perianth and often twisted, contains one cavity with 3 vertical ridges, bearing innumerable tiny dust-like seeds. Most Orchids have storage organs in the form of tubers or swollen underground stems. Some species, for instance, Bird's-nest Orchid (p. 45), are not green, and digest food from decaying plant materials. See pp. 45, 101, 159, 161.

White Helleborine (*Cephalanthera damasonium*)
Broad Helleborine (*Epipactis helleborine*)
Marsh Helleborine (*E. palustris*)
Dark Red Helleborine (*E. atrorubens*)
Violet Helleborine (*E. sessilifolia*)
Autumn Lady's Tresses (*Spiranthes spiralis*)
Creeping Lady's Tresses (*Goodyera repens*)
Common Twayblade (*Listera ovata*)
Lesser Twayblade (*L. cordata*)
Bird's-nest Orchid (*Neottia nidus-avis*)
Coral-root (*Corallorhiza trifida*)
Musk Orchid (*Herminium monorchis*)

Fragrant Orchid (*Gymnadenia conopsea*)
Frog Orchid (*Coeloglossum viride*)
Butterfly Orchid (*Platanthera chlorantha*)
Bee Orchid (*Ophrys apifera*)
Fly Orchid (*O. insectifera*)
Green-winged Orchis (*Orchis morio*)
Early Purple Orchis (*O. mascula*)
Marsh Orchis (*O. strictifolia*)
Heath Spotted Orchis (*O. ericetorum*)
Common Spotted Orchis (*O. fuchsii*)
Man Orchid (*Aceras anthropophorum*)
Pyramidal Orchid (*Anacamptis pyramidalis*)

ECOLOGY OF PLANTS

Ecology is the study of living things in relation to the places where they are found and the factors which influence them. The most important factors influencing the successful growth of plants are (1) soil, (2) water supply, (3) light, (4) climate, especially temperature, and (5) other plants and animals. Each plant has its own particular combination of preferences, though some are more sensitive than others; and these preferences determine the sort of place (habitat) in which the plant can live most successfully. Competition between plants also has an effect on their growth.

In the case of soil, the soil characters exerting most influence on plants are: (1) the available mineral salts, which the plants need for food; (2) the acidity or alkalinity of the soil, to which some plants are extremely sensitive; (3) the presence of salt, which, of course, is only to be found near the sea; (4) the size of the soil particles. Soil in which the particles are very small, such as clay, holds more water than a sandy soil, in which the particles are large.

Plants need a good water supply, but most species cannot grow in completely waterlogged soil because their roots also need air. As plants vary in their tolerance of waterlogging, so they also vary in their tolerance of water shortage. All plants lose water from their leaves and must replace it by taking water from the soil through their roots. Habitats with chalky or sandy soils will obviously have less water in the soil than places such as marshes, bogs, or riverbanks.

All plants need light in order to grow but, as with water, different plants need different amounts of light. Some need relatively little — indeed, may not be able to tolerate bright sunlight; others need full sunlight for successful growth.

Extremes of climatic variation are not found within the British Isles, but the difference in temperature between the north and south is sufficiently marked to affect the distribution of some plants. For example, Mountain Avens (p. 81) grows on mountains in Britain and the rest of Europe and also in the Arctic — all places which are cold or very cold.

Plants compete with each other for space to grow in and for light. Those plants which can reach a piece of open ground first and grow quickly have an advantage over the later arrivals, so plants with widely distributed seeds have a better chance of survival. A well-known example of a successful colonizer is Rosebay Willow-herb (p. 111), which has wind-blown seeds. During the Second World War it rapidly spread into bombed sites, and in peace-time it is one of the first plants to arrive in places such as newly cleared woodland. Competition for light can be seen in hedges where the most successful plants, apart from the shrubs themselves, are those which can climb up to the light by twining (Bindweed, pp. 59, 125), by tendrils (Vetches, p.135), or by hooks or prickles of varying size (Goosegrass, p. 93, and Blackberry and Roses, p. 117).

Man and other animals also affect plants. It is obvious that many of the jobs farmers do are designed to get rid of plants which from his point of view are in the wrong place. Before there were many men in Britain, most of the country was forest-covered, and would be so again if farming ceased. Grazing by domestic animals and rabbits is also important. Its effect can be seen on the South Downs where grazing by sheep and rabbits used to keep the turf short and prevent the growth of shrubs. Now that seedling trees and bushes are not being eaten, much of the grassland is again being covered by Hawthorn and Bramble scrub.

Many plants, especially the common weeds, do not seem to have any special requirements and may be found in almost any place where they can find room, particularly where the ground has been disturbed by cultivation. Other plants are characteristic of certain types of habitat and rarely grow except in these places. Such a preference may be helpful in identification. Some habitats which are worth studying are (1) woodland, (2) grassland, (3) moorland, (4) marshes, riverbanks, and other wet places, (5) the sea coast. Many of the plants growing in these places have some characters which enable them to grow best in their chosen habitat.

WOODLAND

All woods have one thing in common: there is not much light at ground level, especially in summer when the trees are in leaf. Thus most woodland plants are those which do not flourish in strong sunlight, and many of them flower in the early spring, before the tree leaves have cut off too much light. Such early-flowering plants include Dog's Mercury (p. 63), Wood Anemone (p. 67), Wood Dog Violet (p. 165), and Primrose (p. 27). In the more open parts of the wood there are often shrubs, such as Hazel (p. 189), Hawthorn (p. 181), Dogwood (p. 193), and Blackberries (p. 117) as well as the smaller plants and grasses. Many of these woodland flowers are also found in the shade of hedges, but rarely in full sunlight.

The species of plants found in a wood depend to a great extent on the kind of tree which make up the wood and the type of soil present. Some plants, such as Dog's Mercury, are not particular and will be found in most woods.

1. Beechwoods. Beeches do not grow well except on chalky soil. This soil is dry, and the Beech roots run near the surface, to obtain the water before it soaks away. Beech leaves cast a dense shade and they are very slow to rot down, so the conditions for the growth of other plants are poor. Plants which can be found in beechwoods, especially in clearings and on the edge of the wood, include: Sanicle (p. 89), Woodruff (p. 93), and Wild Strawberry (p. 81). Bluebells (p. 169) will often form a sheet of colour in open beechwood glades, and Dog's Mercury is very common. There are some plants which like a soil rich in dead leaves, and so will grow freely in beechwoods. These include some of the Orchids: White Helleborine (p. 101) and Bird's-nest Orchid (p. 45), for example.

2. Ashwoods. These also prefer to grow on limestone and chalk soils. But Ash trees do not cast as much shade as Beeches, so there are often many shrubs present, including Maple (p. 197), Privet (p. 183), and Wayfaring Tree (p. 193). Dog's Mercury and Herb Paris (p. 53) are common ground plants and, if the soil is wet enough, Ramsons (p. 103) and Townhall Clock (p. 49) also flourish.

3. Pedunculate Oakwoods. This is one of the two kinds of British Oakwoods. These Oaks (p. 201) grow on heavy soils, especially on clay, most frequently in the Midlands and southern England. Several other trees often grow among them, including Wild Cherry (p. 195), Birch (p. 201), Maple and Ash (p. 197), Holly (p. 183), Crab Apple (p. 181), and Hornbeam (p. 189). There may also be many shrubs such as Hazel, Hawthorn, Dogwood, and Blackberries, and also Dog Roses (p. 117), Blackthorn (p. 181), and Honeysuckle (p. 15), and, if the soil is wet enough, Willows (p. 187). There is usually a large variety of ground plants in this type of Oakwood. Bluebells, Primroses, Wood Anemones, and Dog's Mercury, as well as Lesser Celandine (p. 5) and Yellow Archangel (p. 29) all grow in large masses, and there may also be scattered plants of Ground Ivy and Bugle (p. 145), Goldilocks (p. 3), Wood-sorrel (p. 85), Barren Strawberry (p. 81), Enchanter's Nightshade (p. 95), and Lords-and-ladies (p. 63), with Woodruff and Sanicle, Wood Sage (p. 31), Foxglove (p. 123), and Bracken on the loamy soils.

4. Durmast Oakwoods. These Oaks (p. 201) are found only on poor, acid soils, either loam or sand, more often in the west and north of England, Wales, and Ireland; but there are some Durmast Oakwoods in the Midlands and south-east England. There are few shrubs, usually Honeysuckle, Blackberries, and Holly, and fewer ground plants. These include Wood Sage and Foxglove, and also Bilberry (p. 121), Golden Rod (p. 33), various Hawkweeds (p. 37), and Bracken.

5. Pinewoods. These are almost always planted, and if the planting is dense, there will probably be no ground plants at all. Pines cast a dense shade, and since they are evergreen, there is no opportunity for the growth of the early spring flowers, such as are found in other woods. The dry Pine needles which fall from the trees do not provide a favourable growing medium for other plants. Where the Pine trees are thinly planted, there may be scattered plants of Bracken, Bilberry, various Heaths (p. 119), and Tormentil (p. 17). Birch trees often grow with Pine on sandy soils, but although they cast very little shade, they grow on such poor soil that few plants will grow with them. The commonest, apart from those found with Pine trees, is Gorse (p. 19).

GRASSLAND

Natural grassland is found only in places where trees and shrubs will not grow, that is, on very shallow soils or above the tree line on mountains. There is much other grassland in Britain only because men and animals have interfered. On land which is grazed by animals or cultivated by man, the seedlings of trees and shrubs are not allowed to develop. If all farming stopped, most of this grassland would, in time, return to forest. The most important plants in grassland are, of course, the grasses themselves; and these, together with the other plants growing with them, vary according to the type of soil.

1. Meadow Land. In meadows, where the grass is cut only once or twice a year, the plants grow fairly tall. Besides the grasses, these plants include Meadow Buttercup (p. 3), Common

Sorrel (p. 59), Ox-eye Daisy and Yarrow (p. 99), and various Clovers (p. 115), with Meadow Rue (p. 5) and Water Avens (p. 117) in the damper places.

2. Chalk Grasslands. These produce low-growing plants because the soil is poor and dry and because the land is usually grazed. Many of the plants growing among the grass develop rosettes of leaves which prevent the grasses from crowding them out. They often flower in early summer before the soil becomes too dry and hot. There are a large number of chalk-loving plants which include: Milkworts (p. 169), several of the Vetches (p. 23), Salad Burnet (p. 157), Sheep's-bit (p. 179), Stemless Thistle (p. 153), Squinancy Wort (p. 93), Rock Rose (p. 15), Knapweeds (p. 155), and Yellow-wort (p. 31). All of these are low-growing, except for their flowering spikes. Certain much rarer plants are also chalk-lovers and can be found only or mainly on chalk grassland. These include certain of the Orchids such as Musk Orchid (p. 45) and Bee and Fly Orchids (p. 161), and the Pasque Flower (p. 139).

3. Other Grasslands. As in chalk grassland, most plants growing in short grass are usually rosette-forming: for instance, Daisy (p. 99), Cat's Ear (p. 37), Ribwort Plantain (p. 61), and Dandelion (p. 41). These perennials are typical of fairly rich soils. On poor, sandy soils, where competition is less acute, more annuals are found, including Eyebright (p. 93), Red Bartsia (p. 141), and Yellow Rattle (p. 25).

MOORLAND AND HEATH

These both occur on poor, acid soils, usually sandy and often covered with a layer of peat. The main difference between them is the amount of water in the soil. The predominant plants are Heaths (p. 119), in particular Ling, which often grows so thickly that not many other plants can compete.

1. Heath. Heaths usually occur in places where there is little rainfall, so the growth of plants is not very vigorous, and there is only a shallow layer of peat overlying the soil. Here the Ling may be mixed with Bell Heather, and there may be also Bracken and Gorse, Broom, and Dwarf Furze (p. 19), as well as smaller plants such as Tormentil (p. 17), Heath Bedstraw (p. 93), Common Speedwell (p. 175), Wood Sage (p. 31), and Bilberry (p. 121).

2. Wet Moorland. Here there is more rainfall and the layer of peat is much thicker and wetter. Cross-leaved Heath grows with the Ling, and where the Ling leaves enough space there are Crowberry, Bilberry, Cowberry (p. 121), and Cloudberry (p. 81), as well as the Wintergreens (p. 85). In very wet places Sundew (p. 83), the Butterworts (p. 139), and Bog Myrtle (p. 191) are found. In the boggy parts the commonest and sometimes the only plant is Cotton Grass, which is not a true Grass but is related to the Sedges. It very rapidly forms peat, which may be up to 30 feet deep.

MARSH, SWAMP, AND WATER

The plants in these habitats vary according to the amount of water and whether it is still or moving.

1. Marsh and Bog. Plants found in marshes and bogs also grow on the banks of streams and in the wet ground round lakes and ponds. In these places the soil is very wet, but the water rises above the soil for only part of the year. There is a large variety of damp-loving plants which can tolerate waterlogged soil which is short of oxygen. As well as the Sedges and some Grasses, these include Marsh Marigold (p. 5), Lesser Spearwort (p. 3), Marsh Ragwort (p. 43), Marsh Pennywort (p. 47), Meadow-sweet (p. 81), Sneezewort (p. 99), Ragged Robin (p. 107), Lady's Smock (p. 113), Marsh Valerian (p. 122), Lousewort (p. 141), Water Mint (p. 143), Marsh Thistle (p. 153), Marsh Violet (p. 165), Water Forget-me-not (p. 173), and Devil's-bit Scabious (p. 179).

Bog is quite distinct from marsh because the water is standing above peat instead of soil, and so is acid instead of slightly alkaline. Only acid-tolerant plants will grow. These include Bog Asphodel (p. 15) and the plants of wetter moorlands.

2. Still Water. In swamps, ponds, and lakes water is standing over the soil all the year round. Swamp plants grow with their roots in the soil below the water, and they have long stems to raise their leaves and flowers to float on the water surface or to rise above it. Some of these have very weak stems and flop down when taken out of the water; this is because, since normally the water supports them, they have no need of the woody tissue which land plants must develop. Again, their root systems are often small because, since they never suffer from water shortage, they need roots only for anchorage and obtaining mineral salts.

Where the water is shallow the plants include Yellow Flag (p. 29), Great Water Dock (p. 58), Water Dropwort (p. 89), Buckbean (p. 95), Gipsywort (p. 97), Water-plantain and Flowering Rush (p. 109), Purple Loosestrife (p. 123), Hemp Agrimony (p. 155), and Brooklime (p. 175). In deeper water can be found Mare's-tail, Water-milfoil, and Horn-wort (p. 53), Yellow Waterlily (p. 7), and White Waterlily and Water Crowfoot (p. 67). Some tiny water plants are not anchored by their roots but are entirely floating. These include Bladderworts (p. 31), Frogbit (p. 103), and the Duckweeds (p. 102).

3. Moving Water. Fewer plants grow in moving water as it is more difficult for them to keep their leaves and flowers above water. Moving-water plants usually have finely-divided or long, narrow leaves, past which the water can easily slide. Plants such as the Pondweeds, Water Starwort, Water-milfoil (p. 53) and Water Crowfoot (p. 67), have many or most of their leaves submerged, and develop more finely-divided leaves in moving water than in still water.

SALT-MARSH AND SEASHORE

The plants growing in these habitats must be able to tolerate a salty soil, either the water-logged soil of the salt-marsh, or the dry, sandy, often shifting soil of the seashore.

1. Salt-marsh. This is a quite distinct type of marsh, most often found round river estuaries. The amount of salt in the soil depends on the frequency with which tidal flooding occurs. The plants to be found are all species which cannot flourish except in a salty soil. They include Rock Samphire (p. 47), Glasswort, Sea Purslane, and Seablite (p. 57), Sea Aster and Sea Lavender (p. 137). Any of these may cover large areas, depending on the amount of silt present and the height above the low-tide level. There may also be found scattered plants of

Sea Wormwood (p. 61), Scurvy-grass (p. 69), Sea Arrowgrass (p. 103), Sea Heath and Sea Spurrey (p. 113), and Sea Milkwort (p. 125).

2. Seashore. On shifting sand the commonest plant is Marram Grass, which is deep-rooting and can rapidly bind the sand.

On more stable sand wetted by spray, Saltwort (p. 57), Sea Rocket (p. 69), and Sea Holly (p. 179) will grow. Further from the sea there is a larger variety, including perennials such as Yellow Stonecrop (p. 15), the Ragworts (p. 43), Hound's-tongue (p. 105), and Creeping Thistle (p. 153). There are also certain annuals which grow and flower early before the soil has dried out. Among these are Whitlow Grass (p. 71), Mouse-ear Chickweeds (p. 75), and Rue-leaved Saxifrage (p. 84).

Shingle beaches are as a rule too exposed and liable to shifting to make it possible for any plants to grow. There are, however, a few special plants of shingly places, such as Yellow Horned Poppy (p. 7), Wild Beet (p. 57), Sea Campion (p. 73), and Sea Pea (p. 135). These plants are perennials, have deep roots, and for the most part are low-growing.

Some plants, though not sea-coast plants, are only to be found within a few miles of the sea. The reason for this preference is not known. Examples are Alexanders and Fennel (p. 47) and Slender Thistle (p. 151).

WASTE AND NEWLY CLEARED GROUND

In areas such as waste places and cultivated land interesting groups of flowers may be found. Newly cleared ground provides a habitat where there is less competition than is normal and where plants with wind-blown seeds can find room to grow. Such land will soon become inhabited by Ragworts (p. 43), Willow-herbs (p. 111), and Thistles (pp. 151-3).

The removal of competition also helps rare plants or plants not native to Britain to become established. A good example is the Oxford Ragwort (p. 43), now very common on waste ground, which was still very rare as recently as 1930. On the other hand, the numbers of some weeds, for example Cornflower (p. 179), Corn Cockle (p. 107), Corn Marigold (p. 39), and Pheasant's Eye (p. 105), have been drastically reduced by more efficient seed cleaning. The activities of man may at any time alter the conditions of any of the habitats described here.

INDEX

Dogwood 192, **193**
Dovedale Moss (Mossy Saxifrage) 84, **85**
Dropwort 80; Water, 88
Drosera 82
Droseraceae 82
Dryas 80
Duckweed, Common, 102
Duke of Argyll's Tea-plant 130, **131**
Dwarf Cornel 192
Dyer's Greenweed 18, **19**
Dyer's Rocket 48

Earthnut (Pignut) 86, **87**
Echium 170
Elder 192, **193**; Dwarf, 192
Elder, Ground, 90, **91**
Eleagnaceae 182
Elecampane 38
Elm, English, Wych, 198, **199**
Elodea 52
Empetraceae 120
Empetrum 120
Enchanter's Nightshade 94, **95**; Alpine, Intermediate, 94
Endymion 168
Epilobium 110
Epipactis 160
Erica 118
Ericaceae 118, 120, 214
Erigeron 154
Erodium 128
Erophila 70
Eryngium 178
Erysimum 8
Euonymus 180
Eupatorium 154
Euphorbia 62
Euphorbiaceae 62, 214
Euphrasia 92
Eyebright 92, **93**

Fagaceae 200, 214
Fagus 200
Fat Hen (White Goosefoot) 54, **55**
Felwort 138, **139**
Fennel 46, **47**
Feverfew, Corn (Scentless Mayweed), 98, **99**
Figwort 140, **141**; Water, 140
Filago 32
Filipendula 80
Fir, Scots (Scots Pine), 184, **185**; Douglas, Silver, 184
Fireweed (Rosebay Willow-herb) 110, **111**
Flag, Yellow 28, **29**
Flax, Fairy (Purging Flax), 82, **83**; Pale, 168, **169**; Common, 168
Fleabane, Common, 38, **39**; Small, 38; Blue (Purple), 154, **155**; Canadian, 154
Fleawort, Field, 42
Flixweed 10, **11**
Flowering Rush 108, **109**
Fluellen, Round-leaved, 24, **25**; Sharp-leaved, 24
Fly Orchid 160, **161**
Foeniculum 46
Fool's Parsley 86, **87**

Fool's Watercress 88, **89**
Forget-me-not, Changing, Common, Early, Water, Wood, 172, **173**; Creeping, 172
Fox and Cubs (Orange Hawkweed) 104, **105**
Foxglove 122, **123**
Fragaria 80
Frangula 182
Frankenia 112
Frankeniaceae 112
Fraxinus 196
Fritillaria 162
Fritillary 162, **163**
Frogbit 102, **103**
Frog Orchid 44, **45**
Fumaria 136
Fumariaceae 30, 82, 136
Fumitory 136, **137**; Ramping, 136; Yellow (Yellow Corydalis), 30, **31**; White Climbing, 82, **83**
Furze, Dwarf, 18, **19**; Needle, 18

Galeobdolon 28
Galeopsis 28, 148
Galium 30, 92
Garlic, Crow, 162, **163**; Field, 162; Mustard (Jack-by-the-hedge), 68, **69**
Gean (Wild Cherry) 194, **195**
Genista 18
Gentian, Autumn (Felwort), Field, 138, **139**; Marsh, Spring, 176, **177**
Gentiana 176
Gentianaceae 30, 124, 138, 176, 215
Gentianella 138
Geraniaceae 128, 168, 211
Geranium 128, 168
Geum 16, 116
Gipsywort 96, **97**
Gladdon 162, **163**
Glasswort 56, **57**
Glaucium 6
Glaux 124
Globe Flower 4, **5**
Gnaphalium 32
Goat's-beard 40, **41**
Golden Rod 32, **33**
Goldilocks 2, **3**
Good King Henry 54, **55**
Goodyera 100
Goosefoot, Red, Stinking, White, 54, **55**
Goosegrass 92, **93**
Gorse 18, **19**; Western 18
Goutweed (Ground Elder) 90, **91**
Grass of Parnassus 82, **83**
Grass Vetchling 132, **133**
Greenweed, Dyer's, 18, **19**; Hairy, 18
Gromwell 94, **95**; Corn, 94; Blue, 170, **171**
Ground Elder 90, **91**
Ground Ivy 144, **145**
Ground-pine 28, **29**
Groundsel, Common, Wood, 42, **43**; Sticky, Stinking, 42
Guelder Rose 192, **193**
Gymnadenia 158

Halimione 56
Haloragaceae 52

Hammarbya 160
Hardheads (Lesser Knapweed) 154, **155**
Harebell 166, **167**
Hare's-foot 114, **115**
Hawkbit, Autumnal, Rough, 36, **37**; Lesser, 36
Hawk's-beard, Beaked, Smooth, 36, **37**
Hawkweed, Few-leaved, Leafy, Mouse-eared, 36, **37**; Orange, 104, **105**
Hawthorn 180, **181**; Midland, 180
Hazel 188, **189**
Heartsease (Wild Pansy) 6, **7**, 164, **165**
Heath, Cross-leaved, Dorset, Sea, 118, **119**; Cornish, Irish, 118
Heather, Bell, Bog (Cross-leaved Heath), 118, **119**; see also Ling
Hedera 64
Hedge-parsley, Knotted, Upright, 90, **91**; Spreading, 90
Helianthemum 14
Heliotrope, Winter, 122
Hellebore, Green, 50, **51**; Stinking, 50
Helleborine, Broad, Dark-red, Marsh, Violet, 160, **161**; Narrow-lipped, 160; White, 100, **101**; Long-leaved, Narrow, 100
Helleborus 50
Hemlock 86, **87**
Hemp Agrimony 154, **155**
Hemp-nettle, Common, Red (Narrow-leaved), 148, **149**; Downy, Large, 28
Henbane 28, **29**
Heracleum 86
Herb Bennet 16, **17**
Herb Christopher 66, **67**
Herb Paris 52, **53**
Herb Robert 128, **129**
Herminium 44
Hieracium 36, 104
Hippocastanaceae 188
Hippocrepis 22
Hippophaë 182
Hippuris 52
Hirschfeldia 8
Hogweed 86, **87**
Holly 182, **183**
Honeysuckle 14, **15**; Fly, 14
Honkenya 76
Hop 64, **65**
Hop Clover (Hop Trefoil) 20, **21**
Horehound, Black, 148, **149**; White, 96, **97**
Hornbeam 188, **189**
Horn-wort 52, **53**
Horse Chestnut 188, **189**
Horse-mint 142, **143**
Horse-radish 68
Horse-shoe Vetch 22, **23**
Hottonia 108
Hound's-tongue 104, **105**; Green, 104
Humulus 64
Hyacinth, Wild (Bluebell), 168, **169**
Hydrocharis 102
Hydrocharitaceae 52, 102, 218
Hydrocotyle 46
Hyoscyamus 28
Hypericaceae 12, 210
Hypericum 12
Hypochaeris 36

PDO 82-1357